Revenge of the Pond Scum:

Searching for the causes of Alzheimer's Disease, Amyotrophic Lateral Sclerosis (ALS) and Parkinson's Disease

by Kenn Amdahl

Books by Kenn Amdahl:

non fiction:

Algebra Unplugged (with co author Jim Loats, Ph.D.)

Calculus for Cats   (with co author Jim Loats, Ph.D.)

Joy Writing: Discover and Develop Your Creative Voice

Revenge of the Pond Scum: Searching for the Causes of Alzheimer's
    Disease, Amyotrophic Lateral Sclerosis (ALS) and Parkinson's Disease

There Are No Electrons: Electronics for Earthlings

The Wordguise Alembic

fiction:

Jumper and the Bones

The Land of Debris and the Home of Alfredo

Revenge of the Pond Scum:

Searching for the causes of Alzheimer's Disease, Amyotrophic Lateral Sclerosis (ALS) and Parkinson's Disease

by Kenn Amdahl

Copyright 2012

all rights reserved

ISBN 978-0-9627815-3-7

Paper version published 2014

Clearwater Publishing Company Inc

PO Box 778

Broomfield, CO 20038 0778

ClearwaterPublishing.com

Wordguise@AOL.com

About This Book

Strange criminals stalk our aging bodies. Millions of lives are destroyed each year by invisible killers who do their evil business and then escape into the night leaving behind sad victims and this mystery: what causes neurological diseases like Amyotrophic Lateral Sclerosis (ALS), Alzheimer's Disease, and Parkinson's Disease? That's the question the author chases throughout this book.

It begins as the journal of a writer (who is best known for his quirky and funny books about electricity and calculus) musing about tropical fish and the difficulties of gardening in Colorado. It morphs into a "nonfiction mystery" about biology and medicine — with twists and turns, suspects and alibis — just like a work of fiction. Unlike fiction, the characters it describes lurk in the shadows of your own life.

When a paralyzing disease strikes one of the writer's friends, he tries to learn about it. He searches the Internet and the library. He reads books, abstracts and articles. He looks up the hard words and tries to make sense of it all. As he learns new information, he records it in a notebook. This book is that notebook. It became both a simplified explanation of science facts and the journal of the author's study of them. It does not follow a perfectly linear path; few adventures do.

At first, he tries to learn about Progressive Muscular Atrophy. He discovers that PMA is related to ALS. As the clues come together, he learns that scientists have good reasons to think ALS, Alzheimer's and Parkinson's are *all* related; things that seem to cause the symptoms of one can also cause the symptoms of the others. In an almost eerie way, some things in the author's daily life also turn out to have a relationship to his study of the diseases.

Ultimately, a combination of factors probably work together to cause these mysterious ailments. The experts don't agree. Some of the contributing factors may be specific toxins, perhaps working in combination. Other factors might include a genetic predisposition, environmental factors, dietary additives or deficiencies of various nutrients. The author describes the clues and leaves any conclusions up to the reader.

Even when the science becomes more complicated, the language remains conversational, with moments of humor. When the author becomes fascinated with things you might think uninteresting or irrelevant, you'll follow right along. Don't worry, he'll veer back before too long. On the other hand, how could someone find the bacteria within pig intestines anything but fascinating? Is there anyone out there who does not love fruit bats? He tracks more than one promising lead only to realize it's a dead end or ancient history. Some of these are the most interesting sections. No matter how well educated you are, you will learn things you probably never even wondered about before.

In a way, that's what this book is about: how the Internet is changing the way amateurs learn technical information. It's becoming easier to find facts; it's also easy to absorb false information. What's hard is telling them apart.

If you've been diagnosed with a neurological diseases, you're probably going to do some research. If your doctor's explanation (and the pamphlet he provides) satisfies your curiosity, this book wasn't written for you. But some people can't help themselves; they will embark on an obsessive quest to learn everything they can from every source imaginable. They may try weird cures from "ancient Egyptian science" or from their buddy's cousin who knew a guy who remembered something his grandmother told him. If that's you, you should keep reading. In your mind, you're a patient hoping to make the best of a bad deal. But many people see you as a faceless potential customer. A few heartless charlatans see you as a mark. They will offer "solutions" designed only to improve their own bank accounts.

This book isn't about any cure, good or bad. Don't expect that. It's just one guy — a nonscientist who happens to write accessible books about difficult subjects — trying to learn about the disease that struck his friend. If you're about to embark on your own research expedition, this book might save you time. It's not a GPS, but it's at least a crude map of one person's path. It's a place to start.

The title probably seems a little strange. As it turns out, many noted scientists think that a chemical produced by the blue-green algae in pond scum can cause the symptoms of ALS. That was one of the first topics that caught the author's fancy. When he was brainstorming about titles for this book with his friend Suzanne (who you will meet on the next page) Kenn jokingly suggested "Revenge of the Pond Scum" and Suzanne laughed out loud. That seemed a good enough endorsement for Kenn.

The cover photo depicts a pond above the Poudre River in Colorado. As far as we know, it contains no pond scum.

Acknowledgments.

Special thanks to the following scientists, professors, and doctors who read and commented on all or part of the book. They corrected many misstatements of fact or focus, as well as errors in my understanding of their theories. The book is much better and more accurate as a result of their kindness, but but their task was not to render it perfect. All remaining mistakes are only the fault of the author. Thanks to:

Paul Alan Cox, Ph. D. (founder and co-director of the Institute for Ethnomedicine)

Cynthia Dormer, Ph.D. (associate professor of nutrition, Metropolitan State University of Denver)

Balz Frei, Ph.D. (director of the Linus Pauling Institute at Oregon State University and expert on exitotoxins and superoxide)

Stephen Lawson, (Linus Pauling Institute Administrative Officer, biographer of Linus Pauling)

Derrick Lonsdale, MD (thiamine expert, Cleveland Clinic, Fellow of the American College of Nutrition (FACN), Fellow of the American College for Advancement in Medicine (FACAM)

James Metcalf, Ph.D (co-director of the Institute for Ethnomedicine, cyanobacteria expert)

Rob Mies, Ph.D. (founder and director of the Organization for Bat Conservation at Cranbrook Institute)

Lisa Miller, (Brookhaven Institute, photographer)

Peter Spencer, Ph.D (director, Oregon Health Science Universities Global Health Centers, expert on BMAA)

John Stanbury, Ph.D. (Founder of ICC-IDD, iodine expert)

Rush Wayne, Ph.D. (author and expert on fungi)

Thank you to the following for the use of their photographs in the printed or ebook versions:

"microcy10" (the image behind the title and author's name in the first ebook) John Patchett (University of Warwick), Mark Schneegurt (Wichita State University), and Cyanosite (http://www-cyanosite.bio.purdue.edu)

"nostoc2c" (the image behind the subtitle on the original e book cover) Roger Burks (University of California at Riverside), Mark Schneegurt (Wichita State University), and Cyanosite (http://www-cyanosite.bio.purdue.edu)

Thanks to Gerry Morrell and the folks at Morrell Printing in Lafayette, Colorado for the picture of Guam on page 18. I saw his photo on a postcard in their lobby, asked if I could use it, and they graciously agreed.

Cycad photo on page 24 by Wikipedia contributor Raul654

Bracken photo on page 35 by Homer Edward Price

Catfish photo on page 40 by Christian Ude

Friends and family who read the book and commented (in no particular order): Suzanne Venino. All of the Amdahls: Cheryl, Paul, Scott, and Joey. Becca Owen, Ruth Coyle, J'nette Plooey, Wade Mayer, Jenni Hadden, Joseph Lawrence Jeannotte, Diane Schumacher, Linda Torpey, and Elizabeth Yarnell. Several gave me suggestions and caught typos; several also gave me encouragement and wine.

Liz Hill gets all the credit for the word "searching" in the subtitle.

Karen Reddick, a friend and very competent editor read the entire manuscript and gently pointed out my most egregious grammar and spelling mistakes. I've added several typos and spelling mistakes since she read it just so it would feel more like one of my books. If you notice mistakes, those are mine, not hers.

Strangers who graciously either gave me information or directed me to other resources:

Betsy Dumont, Ph. D., Gary E. Gibson, Ph.D., Richard L Hanneman, Bryan Haugen, MD, Scott Pederson, Ph.D., Tom O'Shea, Ph.D., Gary Wiles, Ph.D, Michael Zimmermann MD,

And special thanks to all the kind people out there who have read my other books, making the whole writing experiment possible and fun.

I've duplicated this list at the end of the book with website information for many of the people listed.

Contents

Suzanne

Suzanne is a pretty blonde woman a few years younger than me. I've known her for decades and always enjoy seeing her. She's got a big, exuberant personality; she laughs easily and loudly. She loves to ski, hike, and travel. She seeks out new adventures, and everything is an opportunity for adventure: a new kind of food or music or art. Even occasional bouts of poverty between jobs have seemed like a chance for her to try a different lifestyle. My wife tells me that I never have to worry about Suzanne misinterpreting my weird attempts at humor because (she says) Suzanne and I have identical senses of humor. I think that was supposed to be a compliment.

I hadn't seen Suzanne in two years when I learned her company had moved its offices much closer to me. Although we only live a dozen miles apart, she's a busy editor while I'm always busy with small projects and deferred maintenance. When a friendship stretches back through many Presidents, a year or two without sharing lunch doesn't seem significant.

But this time it was. We scheduled a lunch. A week or so before we were to meet, she sent me this e-mail:

"Seeing as we haven't actually talked in so long, you're likely unaware of what's been going on with me. It's a medical thing. I'll give you all the gory details when next we meet, but the short version is that I have been diagnosed with the neurological disease called Progressive Muscular Atrophy. Which is pretty much what it sounds like... my muscles are atrophying because the stupid nerves are dying. They don't know why the stupid nerves are dying. It has mostly been confined to my legs and during the last year I have lost the ability to walk. I use a mobility scooter and am waiting on a souped-up wheelchair as we speak. Needless to say, it sucks. I'm hoping, hoping, hoping they find a cure. That or else I see the image of the Virgin Mary on a taco and am miraculously cured. Could happen..."

I sat back in my chair, stunned, staring at the message on my computer screen. Suzanne in a wheelchair? It was not possible. I checked the date; no, it wasn't April Fool's Day. Suzanne is one of those shiny people, the ones who brighten a room just by entering it. The magic wisecracker who transforms a group of people into a party. I could not even picture her in a wheelchair.

My first reaction was selfish. This wasn't fair to me. I was still adapting to a world without John, another of the shiny people in my life who had died a year earlier in a freak kayak accident. And now Suzanne was sick. It wasn't fair.

Then my inner voice, the stern and sarcastic Keeper of Reason within me spoke: "Well, Kenn," it said. "It's probably even worse for Suzanne."

A medical examiner once told me, after he autopsied someone I loved, that "smart people often grieve by collecting information." I was still grieving for John, but there wasn't any point in trying to collect information about his death. I wasn't grieving for Suzanne. As far as I knew she was sick, not dying. I was just sad and irritated at the universe. But I do process things by gathering information.

So I began researching this disease I'd never heard of, a disease that kills your stupid nerves. The result is the book you're holding in your hands. It's about the disease (which is closely related to Amyotrophic Lateral Sclerosis or ALS), including some startling research about nerve diseases you probably haven't heard about. It's also about Alzheimer's and Parkinson's; it turns out they have many things in common. Finding the cure for one may provide the key to curing all of them, although I did not realize that at first.

Mostly it's about the odd little facts a person learns when he launches himself into an unfamiliar topic. I wanted to be able to have an informed conversation with Suzanne over lunch. I didn't want to write a book. I had no interest in the bizarre topics my curiosity led me to. I had no interest in Micronesian fruit bats, or cyanobacteria, or South American clay deposits, or Oregon pond scum or the mat that holds Middle Eastern desert sands together. I had no urgent interest in Alzheimer's Disease, or autism or Parkinson's Disease or the recipe for 7Up. But one thing led to another and I got sucked in. I spent entire days, then entire weeks reading dull research reports that used many words I had to look up. I learned a lot, and some very impressive and gracious researchers, professors, and doctors read the sections of this book related to their specialties to improve its accuracy (see the "Acknowledgments"). Still, it is possible I did not understand each research report perfectly or explain it precisely. I include a few of my own brainstorms and ideas because I wanted to keep my own mind open and give permission for you to think creatively as well. This is a book about science, but it is not a science book or medical text.

Therefore, do not try to cure yourself based on anything I say. We all know people who believe in a bizarre home remedy we find utterly absurd. We laugh at them and their crystals and magnets, yet we know they're sincere. Every now and then, one of their ideas turns out to be right. I'm usually the guy doing the laughing, but now I know how those people feel. I found myself reading about (and actually considering) some very odd ideas.

I'd like to spark a larger conversation about the things I learned, including a few small details no one but me seems to have found interesting before. Just let me get my aluminum foil hat and we can begin.

Vince and Catfish

I'm pretty handy with a bedpan. And I had pulled wheelchairs up more flights of stairs by the time I was eighteen than many medical professionals will in a lifetime. In a bizarre way, I felt this gave me an advantage over Suzanne's other friends.

Years before I was born, my mother's brother Vince felt sick one night. This was unusual for Vince. He was a lanky Kansas farm boy whose favorite time of year was the wheat harvest. When you worked on a threshing crew, you had to exert yourself to the maximum from before dawn until after dark for days in a row. It was impossibly hard; Vince loved it. Beyond that, he also loved to play football. He wasn't the biggest kid in his high school, but he worked hard. He played on the line where the most physical contests transpire on every play. He scrapped and practiced and finally made it to the first string. The next Friday, for the first time, he would be a starter on the team.

But that night he felt sick. The next morning when he got out of bed, he fell down. His legs simply refused to operate. Polio doesn't care about your plans. Vince lived for sixty-five more years, but he never walked again.

Vince lived with my parents ever since they were married. From the time my sister and I could walk, we fetched things he couldn't reach and pushed his chair over obstacles. Vince baby-sat us when our parents were gone. He taught me chess and the finer points of football and how to repair clocks. Years later, when I was in college, I'd lift him into my red MGB, strap his chair onto the luggage rack, and drive him 500 miles back to Kansas to see his childhood friends. Later still, I helped him buy a house of his own. We argued politics and science, politely but firmly. Until he died at age 83, I was still the person he called when a cat died in his crawl space and needed to be removed. I would complain loudly, employing every vivid word in my prodigious vocabulary. Vince would sit in his wheelchair, coffee cup in hand, eyes twinkling in amusement.

Polio gave Vince one strange gift: for the rest of his life, he never got an upset stomach. No one has ever been able to explain that to me.

For over fifty years, one of the most important people in my life spent his days in a wheelchair. But not his nights. When he dreamed, Vince could always walk. In a way, this is the reverse of most people's experience: when we sleep deeply, our brains usually disconnect from our bodies enough that we can't act out our dreams. It's called "REM atonia." That is, although we don't realize it, we're temporarily paralyzed while we dream just as effectively as if we'd had polio or ALS. Sometimes, this paralysis continues for a while after we awaken and we can't move. It's called "sleep paralysis." Statistically, there's a good chance you'll experience it at least once in your life. If you do, don't panic; it doesn't last long.

I'd seen people's reactions to Vince. Sometimes they spoke in a loud voice, perhaps believing that one's leg muscles turn the wheels that operate one's hearing muscles. Doctors, nurses, and store clerks often spoke to me rather than to him as if he were a child. People's faces showed pity, embarrassment or discomfort; every now and then, they showed revulsion. Sometimes, especially when I was a young man, it made me angry. Sometimes I thought it was funny. Vince ignored it all.

But Vince had decades to adjust to this and Suzanne was still discovering it. Some of her friends were going to treat her differently. New contacts would think of her as "the woman in the wheelchair." Her friends (which are legion) had already organized themselves into a sort of club to provide transportation and help her with shopping. Maybe I could be the friend who didn't notice the wheelchair.

Obviously, I'd have to understand her disease better so I'd know what I was ignoring. Was it fatal? Was it reversible? Was it painful?

I intended to study up on PMA before our lunch, I really did. I got as far as Wikipedia and Google, which only left me more confused. It's related to Amyotrophic Lateral Sclerosis ALS ("Lou Gehrig's Disease") and may be a form of it, but no one knows what causes it. Can that be right, I thought? No one has a plan to cure it? No one really cares because so few people get it? That seemed completely unsatisfying to me. I would have to find a book, or try googling different combinations of words. Surely there was a foundation, and people running marathons to support its cure, and spokesmodels wearing some shade of silk ribbon.

But before I could learn more, I had to deal with my own little emergency. My guppies were dying. It sounds trivial and maybe heartless, but the aquarium sat in my office where I saw it every day; it was my responsibility. Suzanne lived ten miles away and kept busy with her own life. I hadn't seen her in two years. I had to save the guppies. At the time, I had no idea that Suzanne and my tropical fish might be linked in some way.

The tiny universe of my aquarium was being attacked by some mysterious deadly threat. Suzanne fought with a mysterious disease a few miles away. I wanted to understand both threats and even solve them if I could, but I had limited spare time for either. For the next several months, those two conflicts became entwined in my mind. At first, it was just metaphorical, the struggles of one reality mirroring those of another. Later, each world provided clues for the other. Ultimately, Suzanne's disease helped me understand my fish and my fish helped me understand Suzanne's disease.

But at this point, the guppies were my more urgent concern.

I've maintained aquariums on and off for nearly fifty years. In elementary school I raised guppies, monitoring the population in a notebook. By third or fourth grade, when I reached 500 guppies, I started selling them to my school buddies.

Now I don't pay too much attention to my tank but I like having it above my desk. I've reverted to my simple childhood specialty — a few guppies, two albino catfish, a handful of water plants and some duckweed floating on the water.

Duckweed looks like miniature lily pads. Each plant is a quarter of an inch across with a single root descending an inch below the surface. Duckweed reproduces ferociously, guppies love to nibble the roots, and the plants contain more protein than soybeans do. If you forget to feed your guppies, they can survive for a long time on the duckweed.

A year earlier, my duckweed disappeared. No big mystery; I probably just forgot to feed the fish and they ate it all. When I saw some floating at a local pond, I gleefully gathered a spoonful and took it home. Soon my tank looked like the healthy tropical jungle of my elementary school aquariums. Duckweed covered the surface, algae grew like moss on the glass, colorful male guppies chased the much larger females incessantly. The small white catfish scavenged the sand like busy little vacuum cleaners.

One day a female guppy floated lifeless in the water. The next day, two more. One fish might have simply died of old age, but three meant the tank was in trouble. When I looked carefully, I realized that algae covered everything. No longer merely coating the glass, it draped from the other water plants and floated in long strands. This was a good clue.

Like most plants, algae love nitrogen-rich fertilizer. Fish urine contains ammonia, a usable source of nitrogen. When fish create more ammonia than the plants and microbes can use, the concentration can become dangerous. If your plants and algae look lush but the fish are sickly, excess ammonia is one possibility.

I siphoned a few gallons of water from the bottom of the tank, removing waste from the sand in the process, then refilled the tank with clean water. Almost immediately, the fish perked up. I took the bucket of wastewater outside and fertilized some plants with it. Once the water settled, the tank looked just like before, only the fish seemed much more energetic.

I felt quite proud of my tropical fish expertise. The tank thrived.

In fact, a few days later, there were several clusters of eggs on the glass. Guppies don't lay eggs, they give birth to tiny living babies. In all my years of keeping aquariums, this had never happened. My two little white catfish had apparently fallen in love.

I hadn't forgotten about Suzanne's disease and our lunch was less than a week away, but I only had two or three days before the eggs hatched. Once they did, the tiny baby catfish would be eaten within hours by my now enthusiastically healthy guppies. I either had to remove the eggs or relocate all the fish. Either way, I needed to repair another old aquarium, let the silicone cure, set it up, add water, and let it settle for a day. I decided to transfer all my adult fish into the new tank, leaving the eggs undisturbed. With no time to spare, I got busy.

I was going to have to have lunch with Suzanne without really understanding her disease. But man! Catfish eggs! Surely Suzanne would understand.

Lunch with Suzanne

I met Suzanne at a Mexican restaurant. Our mutual friend Cathy joined us, partly because Suzanne could no longer drive. It was good to see them both. Cathy writes science articles for national magazines and was excited about a possible new project. Suzanne is managing editor of an art magazine. The three of us first met years ago because of our interest in writing; we talked about our various writing and editing projects but avoided areas of potential violent disagreement, like serial commas and passive verbs.

Remembering Suzanne's e-mail about her strategy for curing herself — she hoped to see the Virgin Mary on a taco — I had cut out a few images of women's faces from magazines and glued them onto taco shells to give her. One was a picture of the singer, Madonna. One was the actress Jane Lynch (from the TV show Glee) pointing her finger and yelling at someone, and one was a pretty model from a magazine ad for expensive purses. It seemed like a nice assortment of madonnas; surely one would do the trick.

I had mixed feelings about the gag. When someone says they want to see the Madonna on a taco, they mean they're hoping for a miracle. Maybe a friend would not joke about that. On the other hand, sometimes people didn't joke with Vince, because, after all, he was in a wheelchair. That seemed about the dumbest form of prejudice I'd ever encountered. I brought the box of taco shells to our lunch.

Cathy, our mutual friend, loved the tacos. Suzanne laughed and thanked me and was completely gracious. She knew I was trying to cheer her with a little joke. But there was a little sadness in her eyes. She wanted a real miracle, not a gag miracle.

Often, the best gift I can give someone is a wisecrack and a smile. I joke with grocery store clerks and bank tellers and people waiting in the same line at the post office. Nearly always, it feels like I'm doing a good thing, brightening someone's moment with an unexpected giggle.

This time I wasn't so sure. The tacos might have been a mistake.

I asked Suzanne what she knew about the disease. She didn't know much. It seems to be related to ALS, the disease scientist Steven Hawking suffers from. The difference, she said, is that PMA usually does not progress to total paralysis. No one knows what causes it. There is no cure.

We finished lunch. I felt weird enough about the taco joke that I bought the lunch. This was not something a wisecrack was going to fix. Maybe if I had more information it would make more sense, but the subject seemed huge and there were no clues.

As we were leaving, almost as an afterthought, Suzanne mentioned a couple of odd bits of trivia about the disease. For some reason, she said, a lot of people seem to get it on Guam. It's also much more common among soldiers who served in the Gulf War of 1991. As far as she could tell, no one knew if either tidbit was significant.

I felt my brain coming to full attention. Yes, those were odd tidbits. But they were more than that. They were a place to start. They were clues.

By Lisa Miller

Suzanne plants a 75th birthday kiss on the cheek of Jacques-Yves Cousteau
at a party in his honor.

Guam

The word "Guam" conjures a picture in my brain of a tropical island in the Pacific Ocean covered with lush jungle and unpopulated beaches. I associate it with World War II and believe it's an American territory. But I'd never thought much about it. It could be one big volcano or a gigantic airport for all I know.

Similarly, the scene of the Gulf War (Kuwait and Iraq) is, at least in my imagination, one huge hot dry desert populated with rich oil sheiks and poor nomads on camels. I know that's not accurate, but it's also not a lush tropical island.

It seemed bizarre that a disease as mysterious and dangerous as Suzanne's could be prevalent in such different settings. There had to be a clue in there. What did those two places have in common?

Suzanne must have gotten this information somewhere, but no matter how I searched on the Internet, I couldn't really find any evidence to support the idea that PMA actually was more common in those two places. Even when I searched for ALS, I found only passing references. Maybe she had been wrong, maybe this was some sort of unsubstantiated rumor she'd heard. But, in one article about ALS, I saw a reference to a Guam disease similar to ALS called "lytico-bodig." When I searched under that name, I began to understand what she was talking about.

For a hundred years, scientists have tried to solve the mystery of lytico-bodig. It's a weird disease by any standards. Since 1904, scientists have noticed its similarities to ALS: progressive paralysis often leading to death without affecting the victim's intelligence, memories or sensations. In some cases, patients can move in response to stimulations, they just can't instigate actions. For example, a man with the disease might stand perfectly still, unable to move. But, if you toss him a tennis ball, he can catch it. Others become completely paralyzed.

But that's just the "lytico" part of the disease. In some people, it acts a lot like Parkinson's Disease: uncontrollable tremors and twitching. That part of the disease is what the natives call bodig. Many victims get both sets of symptoms, so it is commonly known as lytico-bodig.

What has excited scientists for so long is that the disease only occurs on the island of Guam and two nearby sites. And not to everyone on Guam, just descendants of the indigenous population, the "Chamorros." If it is the same disease as ALS or Parkinson's, or even a very similar disease, then perhaps figuring it out could lead to understanding those others. The more they studied them, the more scientists came to believe they were identical diseases. But ALS and Parkinson's occurred in places like the United States with no apparent pattern. On Guam, they occurred so specifically it was like having the disease under a microscope.

For a hundred years, this has been a mystery with too many clues, too many suspects. And that has driven scientists crazy.

It only seems to affect descendants of the Chamorros; could there be a genetic cause?

It is most prevalent among those who continue some of their ancient traditions; could it be something in their diet? Something in their local medicines?

It hit especially hard in one region; could it be something in the water? Perhaps a mineral imbalance? The climate? Some rare religious practice?

Most tantalizing of all, something about the disease changed over time. To a scientist, change is the Holy Grail. If a pattern changes, then something caused the change. Figure out what caused the change and you're halfway to understanding the whole phenomenon. Consider how this disease changed on Guam:

After being reported in 1904, it remained rare for years. Gradually it became common. By the 1940s and early 1950s it was the number one cause of death among the Chamorro, while not significantly affecting others on the island. That change by itself would have attracted scientists, but the disease wasn't done changing its behavior. In the mid 1950s it began to decline and finally nearly disappeared. In fact, no one born after 1952 has come down with it. But that didn't mean older people were safe. Chamorros who moved away from Guam continued to become afflicted years later. (Many Chamorros prefer to be called "Guamanians." I may use either term.) In many cases, the disease struck twenty years after they left the island. In at least one case, a person first got sick more than forty years after leaving. What kind of disease acts like that? No wonder it was hard to connect cause and effect. Few scientists or doctors were likely to ask a patient what they had for dinner twenty years ago, or where they went swimming, or if their mother gave them any native medicines as a child.

Perhaps the solution was hidden somewhere in the history of Guam. Again, there were too many clues.

In the 1600s Spain colonized the island, bringing Christianity and small pox. Disease and the brutal zeal of the colonizers wiped out nearly all the male Chamorro population. Most Chamorro today have a blended ancestry. If you're looking for a genetic component to a disease, this could be significant.

Before World War II, Japan took over the island. These invaders were equally brutal and the local population scattered into the jungles. Faced with starvation, many were forced to live off the land however they could. Certainly their diet changed during that time. Could that change in diet correlate with the changing nature of the disease? We know the effect might not show up until years after the cause; how long ago should we look for a change in dietary patterns?

When the U.S. evicted the Japanese, Americans appropriated the best beaches for their bases, even though some of this land had been home to the Chamorros. Americans brought guns, candy, and dozens of other things into the island. They brought traditions and perhaps unique diseases themselves.

Could the sudden Americanization of the island be important?

One of the ships that visited Guam during World War II carried stowaways. Tree snakes, six feet long and very hardy, managed to hide in a cargo hold and escape onto the island. Without predators, these nocturnal creatures thrived and multiplied. They ate birds in their nests, they ate bird eggs. They even ate fruit bats, particularly the babies. The slithering reptiles ate whatever small creatures slept in trees. Within a few years, snakes had eliminated all bird life from Guam. No birds whistle or sing on Guam; the lack of bird song remains an eerie aspect of the island. Even the fruit bats became endangered.

But the snakes continued to thrive. By the late 20th century they routinely gathered in places they should have avoided, causing electrical power outages.

Could the tree snakes have something to do with it? Obviously, cases had been reported decades before the arrival of the snakes, so they didn't cause the disease. But could the slithering population explosion have some connection to its changing nature?

A friend of mine went to high school with a Chamorrita (that's the feminine version of Chamorro) who had moved to the United States. The Chamorrita teenager hated to drink plain water. Always had to be soft drinks. Well, many teenagers are like that, it's probably nothing. I didn't follow through on the tip, but I couldn't help wondering if I was ignoring a clue. That's the problem: you can't tell what's a clue and what's just a weird fact.

Different scientists became enamored with various theories and spent lifetimes trying to prove them. One group tested the drinking water of a Chamorro village and discovered it to be very low on calcium and magnesium. Drinking that water could, they reasoned, lead to an excess of aluminum in a human and aluminum was already a suspect in some other neurological diseases. They threw themselves into research along these different lines — genetics, minerals in the water, history — with missionary zeal. Some believed they had the solution, they published papers, they held private celebrations. Each time, a flaw in the science, a flaw in the reasoning, or someone else with a more compelling idea ultimately discredited them. In some cases, they had spent entire careers in this search and, when their theory failed to withstand criticism, they were devastated.

These detectives concocted many theories, but they couldn't completely ignore one suspect, despite its perfect alibi. For a hundred years it had been considered, studied, proven innocent and released, only to be brought in for questioning once more, a decade later, when new detectives took over the case. This particular suspect looked so harmless, so benign — even pretty — you felt guilty for even considering it. Yet it always seemed to be near the scene of the crime. It was a primitive tree-like plant that flourished when dinosaurs owned the world, and the Chamorro ate its seeds.

It's called the cycad tree. I wanted to learn more about it, but first I had to deal with my aquarium.

Catfish Eggs Hatch

The catfish eggs disappeared from the sides of my aquarium overnight, but I couldn't see any baby catfish. Tiny white fish (smaller than a mosquito when they hatch) easily disappear against the white gravel. Even if they didn't, newly hatched catfish quickly hide, then live for a week or so on the egg sack they're born with. No need to move, no need to do anything except get used to the strange new world around them. There's no point feeding them during this time; they won't move much or eat, which makes them nearly impossible to spot. The proud human stepfather can only peer into the tank and wait.

In a week or so, I began to see random specks of white dust, too small to be sure of, dart an inch or so up from the sand before vanishing again. I started feeding some powdered brine shrimp. Within another week they had grown into busy little toddlers, each the size of a gnat, merrily bustling around on the sand. My childhood habit of documenting the flock resurfaced; I counted 56 babies in there.

For all the fish I kept as a kid and all the live-bearers (like guppies) I raised, I didn't get my black belt in tropical fish: I never bred egg layers. All the egg layers seemed to have exotic patterns to their reproduction and raising them was a tricky business and expensive. I spent hours reading books about them; I learned their bizarre rituals, their unique water requirements, their special dietary needs. But breeding egg layers felt like something grown ups did.

I say all this so you can appreciate the excitement I felt at seeing 56 baby catfish, hatched from eggs, scurrying around the bottom of the tank. I knew dumb luck had put a male and female catfish in my tank. I hadn't tried to guess their sex and the kid in the shopping center pet store probably didn't even know they came in sexes. I hadn't fed them a special diet or checked the pH of the water. My only contribution was recognizing the eggs and evacuating the rest of the fish to another tank, newly set up for them. But I could now say I had bred catfish. I was very excited.

A few days later my excitement turned to anguish when the babies started to die.

Baby fish are fragile. The vast majority never survive to adulthood. And with 56 gnat-sized white fish intermittently scurrying around and stopping on white sand it's hard to tell how drastic the situation really is. Are they hiding? Are they sick? Or are they sleeping?

Within a few days I knew the tank was in trouble. I saw two actual dead fish and very few active ones. The fish I could see acted lethargic. I wasn't equipped to test the water, but I had some good clues. The sand looked dirty below the surface, even though I'd cleaned a lot of it not long before. The duckweed was growing like crazy and so was the algae that coated the back and sides of the tank and even seemed to coat the other plants. This was just like before. I thought I had cleaned enough of the waste material out of the sand when my guppies were dying in this tank, but obviously I had not. There must still be too much ammonia in there. I needed to clean the sand again and replace half the water. But that was much harder when tiny catfish hid in that sand.

Being as careful as I could be not to suck up any living catfish, I siphoned water from the bottom of the tank. I've done this hundreds of times in my life. The water was dark, with a distinct fishy smell, but not anywhere near as bad as some I've seen. As is my custom, I poured some of the nutrient-rich waste into my terrarium to fertilize the tiny eucalyptus tree, miniature bamboo, and aloe plants. Coincidentally, a neighbor had become bored with his two fire bellied toads and had donated them to the terrarium. I wanted to keep it lush for its new occupants.

I had already bought a number of tiny crickets for my new toads, but amphibians also love earthworms and those contribute calcium to their diet. The coffee can full of dirt and earthworms was starting to look dry, so I added a little nutrient water to it as well. The rest I lugged outside and dumped on a corner of my garden. I felt like quite the organic gardener.

A few days later, the baby catfish started dying off again. This called for extreme measures: I decided to ask someone for advice. I took a sample of the water to a local pet store for testing. The guy who did the testing was a little older than most of the staff. He knew a lot about fish. He tested everything he could test. When he was done, he smiled at me.

"That's some of the best water I've tested in a long time," he said. "Nitrates, perfect. Nitrites, perfect, pH perfect. You must know a lot about aquariums."

"Yeah," I said. "I just don't know why my fish are dying."

26

Maybe my subconscious would solve the catfish problem while I slept. To give that part of my brain space to work, I needed to think about something else for a while. I went back to studying clues about Guam and the mysterious lytico-bodig disease. Scientists who studied it seemed very interested in this plant I'd never heard of, the cycad tree.

The Cycad Tree

The cycad tree isn't really a tree at all. Nor is it a palm or a fern, although it looks a little like them. It's a cycadophyte, an ancient type of plant only distantly related to the ones it resembles. It's got a thick trunk, topped with palm-like fronds. The entire plant isn't huge, by tree standards, only growing fifteen or twenty feet tall (the "giant cycad" of Thailand reaches heights of 25 feet). The fat, rough trunk is thick as a telephone pole.

Biologically speaking, cycads are about as closely related to trees as you are to a scorpion. They evolved three hundred million years ago, which is about a hundred million years before the first flowering plants. To put this in perspective, the term "flowering plants" includes nearly every variety of land plant you or I could name: grasses, trees, weeds, vegetables.

Although they are relatively rare today, for a few million years, cycads flourished. Without many animals to consume them, they spread gleefully across the planet. Back in the Carboniferous era, plants similar to modern cycads (as well as ferns and other ancient relatives) died and piled up and were covered with sediment. Gradually they became coal. The United States alone mines a billion tons of coal every year; those ancient plants were very prolific. The exact age of the cycad remains controversial.

As the Carboniferous era ended about 300 million years ago, cycads started taking over. For the seventy million years or so after that, cycads ruled. Then the dinosaurs showed up to nibble on them and they had to develop new coping strategies or die. By that time, cycads were probably the dominant plant on earth.

They survived the munching of dinosaurs, climate change and the rise of birds and mammals. Their descendants, largely unchanged, continue to live today in pockets around the globe much as they did before the first flower bloomed on earth, before the first amphibian crawled or the first insect buzzed.

Cycads can thrive in poor soils where more "advanced" newfangled plants shrivel and die. This is because millions of years ago they made a bargain with a variety of bacteria called "cyanobacteria." In exchange for a moist, secure home within the plant's roots, the bacteria fed the cycad. The bacteria fixed nitrogen into ammonia. They used some of it themselves and the excess fed the plant. With their own nutrient supply, the cycad tree can grow in rocky soil that would starve another plant. And the cyanobacteria can survive in a climate that would otherwise kill them. This kind of symbiosis isn't uncommon in nature, although we usually think of it occurring in the roots of legumes like peas and beans.

The cycad tree does not grow everywhere on earth, but it's common on Guam. It produces a nut that is both tasty and nutritious. Animals eat it, humans eat it, even fruit bats eat it. The only drawback is that it's poisonous. Most interesting of all, at least some of the poison isn't manufactured by the cycad plant itself: it's manufactured by the cyanobacteria living in its roots. In addition to fixing nitrogen for the cycad, the bacteria produce toxins that spread through the plant, making it much less attractive for animals to eat. The poison isn't so concentrated that it kills most animals after a few nibbles, but eating the cycad is a bad long-term strategy.

Cattle that eat cycad leaves or nuts can become paralyzed and even die. In Australia, this condition is called "zamia staggers." The leaves of many kinds of cycads contain a chemical called cycasin. Intestinal bacteria break down cycasin, leaving "methylazoxymethanol" or "MAM," a toxin that causes a degeneration of the spinal cord. It can take a year or more after eating the stuff for the results to show up. Cycasin can also damage chromosomes and lead to cancer. We've known about this toxin since about 1860.

Guamanians love the cycad nut. They make it into a flour, they make bread from it, they cook it like oatmeal. They know it's toxic, but they also know they can eliminate the toxin by slicing the nut into thin sections and soaking it for a few days before they use it.

You can see why the cycad tree would intrigue scientists studying lytico-bodig. It's a toxic plant used as a food by the native population. If you were searching for a unique dietary factor to explain a disease that struck Guamanians, you couldn't resist the cycad nut. And so generation after generation of scientists studied it.

One particular toxin from the seeds proved especially tantalizing. Beta methyl amino alanine, abbreviated BMAA, is found in the cycad nut in small amounts. It's actually a kind of amino acid that isn't normally used by humans. BMAA is one of the toxins produced by the cyanobacteria in the plant's root system. Suspiciously, it was found in the brains of people who died from lytico-bodig. On the other hand, they were Guamanians. Of course they ate cycad nuts, of course they would have trace amounts of this chemical in their bodies. It seemed like such a good suspect that one scientist, E.A. Bell, experimented by giving it to rats. It didn't seem to affect them. In 1971 he reported they did not show symptoms after feeding them BMAA for 78 days. The BMAA hypothesis faded.

Ten years later, well-respected researcher Peter Spencer took it up again. He approached Peter Nunn, who had worked with Bell, and got some BMAA from him. When they fed it to rats, nothing happened. But, when they put some on individual rat brain cells, it killed the cells. Beginning in 1985, they fed it to monkeys and reported on it in 1987. In small doses, it didn't do anything. But, in big doses over a month or two, it did in fact cause the exact symptoms of lytico-bodig in the monkeys.

The burst of excitement about this finding was short lived. Yes, BMAA could cause symptoms that mimicked the disease, but only when administered in huge doses. Critics pointed out the fantastic quantities of cycad nuts a person needed to consume to get these dosages in their diet. It simply wasn't possible. In 1990, Spencer published a report agreeing that the BMAA hypothesis fell short of proving it caused the disease in humans.

One can imagine a scientist spending years of his life trying to prove a hypothesis, then having his experiment succeed validating all those years. One can also image the bitter disappointment of realizing later that, no matter how you look at it, the dosages required were simply too high. All those years actually proved you were wrong. You chased a lovely obsession that finally broke your heart.

At least that's how I imagined it and how some of the people who wrote about this described it. After I wrote the first draft of this book, I tracked down Peter Spencer to make sure I represented him accurately (which became a lot easier after I stopped spelling his name "Paul Spenser"). He said no, I did not have it quite right. He said, "Of course, the primates findings were and continue to be exciting, but the primate model of L-BMAA toxicity was never "sold" as an accurate model of Guam ALS-PDC. All we said was that the animals showed clinical and neuropathological changes that one would like to see in a model of the human disease but that cessation of L-BMAA dosing led to disease recovery. This is the point you misrepresent: scientists such as myself are our own worse critics: we strive not to go beyond the facts at hand and get upset when the rest of the world proceeds unfettered along that course. So, I am sorry, I never had a broken heart in the way you describe."

OK, so I would have had a broken heart, but Dr. Spencer did not. I wish I'd spelled his name correctly in my notes, because he's done a ton of work in this area since then and knowing about it could have saved me a lot of time. The BBC produced a documentary about his work on BMAA called "The Poison that Waits" which you can watch online. I recommend it. He seems like a neat guy. He has sent me some great information and I hope we can meet some day. He kindly pointed out that lytico-bodig also occurs in two other locations in the South Pacific: Kii Hanto, Honshu Island, Japan and in Papuan New Guineans living in West Papua, Indonesia. Therefore, he's careful to refer to the disease as "western Pacific ALS-PDC" not "Guam ALS-PDC." As it turns out, this could actually be quite important in a later chapter.

But, at this point in my story, many people considered cycad nuts and BMAA an outdated theory, like the flat world theory. It was simply wrong.

Unless you thought of it differently. Spencer's experiment didn't provide the solution and he never claimed it did. But maybe it provided a clue.

In 1996, the famous writer Oliver Sacks wrote two related books that talked about lytico-bodig, *Island of the Colorblind* and *Cycad Island*. These were published in one volume and explained a lot about the diseases. Mr. Sacks seemed to sympathize with Spencer's idea even while agreeing that it had not been proved. By the time I got interested in this project, that volume had become relatively unknown and did not show up on the first pages of any Internet searches I ran. Had I discovered these books early on, they would have also saved me a lot of time and might have convinced me not to dig any deeper. They are well written and provide a much more human picture of the diseases than one gets by reading scientific abstracts online and more than you get by reading what I've written here. When I discovered the book and ordered it, I was afraid a better writer had already done my little project. I recommend his book, but we've learned a lot since he wrote it. I'm glad I didn't read it until I was nearly done with my own book.

Similar Diseases

When you use the Internet to research something you don't understand, you use the few "keywords" that you do know. This can lead to interesting new information, but will also lead down a series of dead ends. Searching "lytico-bodig" leads to articles that mention "cycad trees." Searching "cycad tree" might lead to an article containing the word "Chamorro" which might lead to an article about the native diet on Guam that mentions kabocha squash. "Kabocha squash" sounds tasty, but has nothing at all to do with what we're trying to learn. I followed a whole bunch of dead ends like that, learning about things that have no place in this book. Mostly I've spared you from those irrelevant details.

Two weeks after reading about kabocha squash (also known as Japanese pumpkin) and determining it was just a weird little side trip for my brain, I saw some kabocha squash for sale in the local Asian grocery store for the first time. I felt compelled to buy one. It was excellent baked with cinnamon, more a dessert than a vegetable. But still irrelevant.

A different path led a more interesting direction. ALS and PMA are neuromuscular diseases. Other than my uncle Vince's polio, I knew nothing of diseases that damage or destroy nerves and therefore renders muscles useless. So I started a haphazard search of nerve diseases and paralysis in humans and animals.

Most of the time, people become paralyzed for reasons that are obviously unrelated to ALS, like car wrecks. But sometimes I'd read a sentence and instantly become more alert. I'd think, "that might be important. That could be a clue." I felt like Sherlock Holmes, noticing the smudge on the mirror the police had missed. Only in my case, more educated minds had long since concluded their investigation and, in most cases, the culprit had been caught and convicted. I became more interested as I started to notice some patterns. The excitement of discovery was real; I just arrived on the scene a hundred years too late.

Or, in the case of lathyrism, two thousand years too late.

The symptoms of lathyrism are eerily similar to PMA — paralysis of the lower body, caused by nerve damage to the "lower nerves." The "lower nerves" are the ones extending from the spinal cord to the arms, legs, and rest of the body. The brain and spinal cord are "upper nerves." We've known about lathyrism since the very first doctors: Hippocrates wrote about it. It continues to affect thousands of people today. Yet it is completely preventable. We know exactly what causes it.

You get lathyrism by eating too many grass peas. To prevent the disease, people could simply stop eating them. But, when you're starving, you take the risk.

Like other legumes, grass peas develop a symbiotic relationship with certain nitrogen fixing bacteria. The bacteria live in little swollen areas of the root and convert nitrogen in the air or soil into plant food. Because of this, peas can thrive in terrible soil where most self-respecting vegetables would despair and die.

This sounded eerily similar to the survival strategy of the cycad trees. I snapped to attention. Then my sarcastic inner voice started sneering at me. When you're looking for clues, everything is a clue. Internet searches that include related words will probably always lead to websites with similarities. Besides, I reminded myself, the cycad tree theory had been discredited.

The grass pea is toxic, but human bodies can process the toxin over time. You're not likely to become paralyzed the first time you eat it. In times of drought, poor people in India and elsewhere turn to eating the toxic grass pea. It can keep them alive until the drought ends, assuming the rain arrives soon enough and other food becomes available. But, if grass peas are your main diet for a month or so, you are very likely to be paralyzed for the rest of your life. This cruel bargain has been destroying lives for more than 20 centuries.

We think of legumes like peas and beans harboring nitrogen fixing bacteria in their roots. But hundreds of other kinds of plants do as well. For example, alder tree roots contain bacteria from the "frankia" group (rather than the cyanobacteria crowd) and therefore these trees can survive in poor soil. Interestingly, alder wood is considered "toxic" and woodworkers are encouraged to protect themselves from inhaling its sawdust. The rare "seaside alder" contains potent poisons that a few scientists have been examining as possible medicines. This isn't as weird as it sounds; for a hundred years we've been treating cancer with poisons that are only somewhat more dangerous to cancer cells than they are to healthy cells. Learning about grass peas and alder trees felt like progress. It turns out there are many more plants that cause the same kinds of symptoms.

Peter Spencer, the guy who fed BMAA to monkeys, studied lathyrism before he studied cycads. It turns out the poison in grass peas resembles BMAA in many ways. It's called BOAA.

I kept discovering similar stories and they all went like this: Some plant maintains a symbiotic relationship with bacteria. The bacteria fix nitrogen for the plant. The plant is useful to man for some reason or another, but it is also poisonous. If we eat it, we develop symptoms of one of the terrible neurological diseases like ALS. In most cases, I could not find evidence that people believe the bacteria associated with the plant manufacture the toxin that causes the disease. But the same general story kept repeating itself so often that I found myself looking for examples everywhere. I found myself wondering if all the toxins produced by every poisonous plant in the world comes from some bacteria working in secret cooperation with it. I confess that I still harbor that suspicion, although I have absolutely no evidence to support my superstition. On the other hand, you've got lucky golf shoes so be careful when you mock me.

We know of many odd couples in nature working symbiotically. The pairs can include bacteria, fungi, algae, and various members of the animal kingdom. In some cases, the bacteria or algae's primary function is to produce extra oxygen. In some cases it consumes waste products so the animal doesn't succumb to it's own poisons. Sometimes algae takes up residence within an animal to provide extra carbohydrates via photosynthesis. We've known about this for decades.

In his 1924 book *Symbiogenesis: A New Principle of Evolution* by Boris Kozo-Polyansky, (which Harvard University Press re-released in 2010) the author mentions the "classic dark green worm" *Convoluta roscofensis*. For some reason, the worm's scientific name has now been changed to the snappier *Symsagittifera roscoffensis*. These little fellows live in shallow water. Algae within each worm make it green and serve as an internal solar food farm. The book refers to a scientist named Graff demonstrating in 1891 that mature worms in good light live for weeks without any food by surviving on the food generated by the algae's photosynthesis. Worms kept in the dark die of starvation.

One of the weirdest examples of symbiosis is between cyanobacteria and sloths. Apparently, the bacteria set up residence in the fur of the sloth. Because of their coloring, the blue-green algae provide camouflage for the slow moving mammals. In return, the bacteria receive a nice warm place to live plus free, although not speedy, transportation.

We also know that the toxin "tetrodotoxin" is manufactured by a few different bacteria (for example, *Pseudoalteromonas tetraodonis*) that live symbiotically within different kinds of animals. The most familiar host is probably the puffer fish, but the bacteria and poison are also found in an octopus and in some amphibians. The puffer fish provides a nice environment for the bacteria. To pay its rent, the bacteria make the fish deadly poisonous, which discourages predators. Tetrodotoxin achieved some notoriety when Wade Davis of Harvard (now of National Geographic) described its use in voodoo to create "zombies" in his book *The Serpent and the Rainbow*. Its use by insects is related in Thomas Eisner's wonderful book, *For Love of Insects*.

Lynn Margulis, who was involved in the re-release of *Symbiogenesis*, has promoted the idea of "endosymbiosis" for decades. In this theory, evolution was more about cooperation than competition. She suggested that primitive one-celled life forms like bacteria invaded cells in a sort of symbiosis that became permanent. She was ridiculed for decades until studies revealed that the DNA in mitochondria and chloroplasts does not match the DNA in a cell's nucleus. Now her idea has become nearly universally accepted. One should not discredit an idea just because established scientists can't wrap their brains around it at first.

The point is this: bacteria live symbiotically within some animals; sometimes their job is to manufacture poisons. The poisons protect the animal from getting eaten. Bacteria also live symbiotically within some plants. When their job is fixing nitrogen to feed the plant, humans get interested. We invest lavishly to study those relationships because bacteria are a lot cheaper than fertilizer. We're less interested in any other roles bacteria might perform within their host plant and therefore we don't know as much about them. One of those tasks might be creating distasteful and deadly chemicals. Poisonous plants don't get eaten as often, so that symbiosis could be a win-win deal. The cyanobacteria fixing nitrogen for the cycad tree also manufacturing poisons. One of these poisons has been found in the brains of people with lytico-bodig and also in the brains of people who died of Alzheimer's. That's why we're interested in them. Maybe other plants maintain a similar relationship with bacteria that don't happen to fix nitrogen.

As I continued along, I perked up whenever a poisonous plant had a big root, or a horizontal rhizome root because those seemed friendly places for a nice bacterial infestation. I also perked up whenever I stumbled across a plant that was poisonous if we had reason to believe it might be hosting some variety of bacteria. Especially cyanobacteria. Mostly, I found no evidence whatsoever that my little superstition had any basis in reality, but the idea did start to influence my thinking so it's only fair for you to know it.

Once you start looking for poisons that come from plants, especially "toxins that paralyze" produced by plants that might harbor bacteria in their roots, you discover an awful lot of interesting plants and poisons. The plants that produce toxins include many that are also useful as foods or in other ways. You also discover many toxins that humans use as insecticides, herbicides, fungicides, and in manufacturing processes.

Cassava is a hugely important food around the world. The root is as thick as a man's arm with the consistency of a potato and rich in starch. In many countries, cassava is the most important food crop, or one of the top few. Some varieties are bitter because they contain cyanide. Those varieties are often preferred by poor farmers because the bitter taste thwarts insects and animals. If you soak the root in water, then boil it in water (and remember to throw out the water you boiled it in) you eliminate the toxin. If you don't process bitter cassava root, you get a disease that is very similar to ALS called konzo. In times of drought or war, people don't always process the root correctly and the result is a lot of paralyzed citizens.

Did cassava have nitrogen fixing bacteria in its root system? No, it did not appear to. What it did have was a symbiotic relationship with fungus around its root system. These fungi are called "mycorrhizal fungus." Interesting, I thought. As it turns out, something like 80 percent of all vascular plants have this sort of symbiosis with fungus. Vascular plants have "vessels" like our blood vessels that carry water from one area to another. Algae are nonvascular, while trees are vascular. The fungus helps provide nutrients for the plant in exchange for the products of photosynthesis, like carbohydrates. People who own tree farms inoculate the soil around new trees with some of these fungus. If they don't, the trees don't thrive. One kind of mycorrhizal fungus is truffles. Another is the deadly amanita mushroom. I made a note to learn more about those at some point and moved on to other plants that cause symptoms similar to ALS or PMA.

Bracken is a lovely fern that thrives in patches of forest cleared by loggers or fire. It stands about two feet tall. Like the cycad, bracken was an ancient plant when dinosaurs walked the earth. They don't flower or produce true seeds. Instead, hundreds of tiny round spore cling to the undersides of each leaf. Each of these tiny little spheres contain half the complement of chromosomes of the plant, much like a sperm or egg does in animals. They fall to the earth and begin to grow slowly. When a pair find each other, they join their DNA together and grow into a new plant. In this way, they act like mushrooms, but it's a superficial resemblance.

Rather than traditional roots that aim downward, the underground portion of a bracken is a thick horizontal rhizome. These rhizomes store energy like a carrot or potato does, allowing bracken to spring back after a forest fire much faster than other plants. These rhizomes can spread underground, and new plants can grow from them. Like other vascular plants, many kinds of ferns have a symbiotic relationship with fungus around their root systems, but I couldn't tell for sure if bracken does. Bracken ferns hinder reforestation. After a fire, they pop up first and shade the ground. They prevent tree seeds from getting the light, warmth, and moisture they need to sprout. Lovely and ancient though the plants may be, bracken is considered one of the most invasive pests on the planet.

When horses or sheep eat bracken, they lose control of their muscles and can no longer walk. The symptoms are very much like ALS. If they keep eating it, the disease often kills them. Animals don't eat bracken if there's anything else to eat. This gives the fern an advantage over other plants. Sometimes horses eat pieces of bracken mixed accidentally with their hay. Interestingly, cattle can tolerate a lot more bracken than horses can. But if they eat enough, cattle develop the same symptoms.

People aren't as smart as horses; they often eat bracken, especially the new growth in the center called a "fiddle head." In some parts of the world, it is considered a great delicacy. People know the plant is toxic, but they also know they can eliminate the toxin by soaking the plant in water for several days and then discarding the water. Cooking seems to detoxify it as well.

I had seen bracken fern many times in my youth. It looks like an ancient plant from the jungles of the dinosaurs, so I'd always wanted to grow some in a terrarium. What better background for my salamanders, frogs, toads, anoles or turtles? Whenever my family visited an area where it grew, I quietly slipped a couple of leaves loaded with spores into my shirt pocket. But they never sprouted in the terrarium. As a kid I hadn't studied their life cycle; I didn't even know what the plant was called. This year, after reading about them and how they spring back after a forest fire, I decided their roots were the key. While I was in Oregon, I dug up a little bracken fern and a whole bunch of root. It is now thriving in a terrarium.

One more little lifetime goal checked off my list. It transplants wonderfully, if you get enough of the root system. It looks really cool in there, but I'll resist the temptation to add some to my salad.

Bracken's rhizome root system is a big, fat, juicy potential wonderland for microbes. The rhizome is the most poisonous part of the plant. I wondered if maybe bacteria lived down there? I could find no evidence, so I can't suggest that as a possibility. But, knowing my little superstition, you won't be surprised that I retained a flicker of hope that someday someone would discover that bacteria produce whatever the bracken toxin was. Or that they already had and I'd be able to find their report. But it eluded me for now.

Months later, I learned why bracken is so deadly. I could just tell you now but, in order to maintain the sense of the order in which I learned things, I've added that in a later chapter. Anyway, at this point in the story, I stumbled onto something that seemed even more interesting.

In reading about bracken, this pesky fern that takes over forests and sickens humans, horses, and sheep with symptoms very much like ALS, I found an awful lot of references to pigs.

Bracken photo by
Homer Edward Price

Pigs

Pigs just love bracken. They dig up the rhizomes and eat them like candy. The fact that the plant is poisonous doesn't seem to bother them at all. We've known this for a long time, but there hasn't been much practical use for that information. Until 2004.

After two thousand years of man's extracting resources from it, the Caledonian Forest in the highlands of Scotland has become a shadow of its original glory. Once covering over 5,000 square miles, trees now cover less than one percent of that. Part of the problem is that, once you cut down the old growth, it's hard for trees to compete with bracken. The bracken plants spread, they grow, they choke out the trees and wildflowers. Natural browsers like deer leave them alone.

Then Liz Balharry and Rae Grant got the idea to import a few wild boars to eat the bracken. Once common in the Caledonian Forest, wild boar had been hunted into extinction hundreds of years ago. Could reintroducing them help solve the reforestation problem? Liz and Rae started the Guisachan Wild Boar Project near Tomich to answer the question in a scientific way. They brought in some partners, including Trees For Life and the Forestry Commission of Scotland, built enclosures so the boars wouldn't just run loose, and imported a few boars. For three years they monitored the experiment.

The wild pigs did exactly what Liz and Rae hoped. They dug up the bracken rhizomes and ate them. Tree seeds and wildflowers sprouted in the naturally plowed ground left behind. When new bracken sprouted, the pigs ate them too. Sure, they also ate some of the new saplings and chewed the bark off a few trees. But they proved the concept worked. In 2009, Trees For Life built a 30 acre enclosure on an estate in the forest called Dundreggan and acquired six wild boar from the Royal Zoological Society to expand the research. By early 2010, two of these boars had died of unrelated disease, but the project continues. Their website reports that robins follow the wild pigs around, waiting for them to root in the dirt for rhizomes, because that exposes earthworms. Certainly the early bird gets the worm, but the smart bird lets a big pig do his digging for him.

Genetically, pigs are closely related to humans. We even use pig tissue in some human operations. Yet the stuff in bracken that can kill horses, cows, and humans doesn't seem to bother pigs nearly as much. It makes you wonder what pigs have that we don't have.

Bracken isn't the only poison pigs seem immune to. They also love to eat rattlesnakes. Pigs will race past much easier food to stomp a rattlesnake to death and enthusiastically gobble it down. Farmers and ranchers have known this for generations. When the rattlesnake population gets out of hand, ranchers bring in the pigs to kill them off.

Never having maintained any porcine relationships, I found this a startling tidbit. Even odder, experts don't seem to agree on why the pigs seem completely unfazed by the venom. Some maintain that pig skin is too thick for the fangs to really penetrate. Perhaps that's all there is to it, but I'm surprised I could not find any conclusive evidence of that in the two minutes I spent researching it. So I looked for other explanations as well.

We know that some bacteria digest various kinds of toxin. That's why all the toxins ever created on Earth aren't still here — bacteria eat them. Could pigs harbor a bacteria that protects them? In 1979, an Australian scientist named Edward G. Russell examined the bacteria in a swine's large intestine. Of the 192 varieties he observed, only 124 could be identified by species. Sixty eight kinds of bacteria could not be identified at all. By 2001, in her doctoral dissertation about the "Immune Activation of Swine Gastrointestinal Epithelial Cells in Response to Microbial Exposure" Kristine A. Skjolaas says that "the gastrointestinal tract is home to roughly 500 to 1000 species of bacteria." First, Kristine, forgive me for not reading your entire paper, it's just way above my head. What I could understand was fascinating. Right now what's interesting is that in 22 years it appears we've discovered a whole bunch more bacteria inside the average pig gut. But even at that, if the current range is somewhere between 500 and 1,000 species, our margin of error is 500 varieties. We clearly have not exhausted our investigation. There must be somewhere between 68 and 500 species of bacteria we have not even identified; one of them might love to eat bracken toxin and rattlesnake venom.

Now I understand bracken a whole lot better and it's more interesting than I imagined when I started. But when I was researching this, my brain just kept going over all the possibilities. Here are some of the notes I wrote at the time:

Much of our immune activity transpires within our intestines; maybe intestinal bacteria manufacture substances that counteract the toxins.

Or maybe pigs create more of some enzyme that detoxifies the venom. Or might something in the piggish lifestyle counteract the poison?

Would the answer to this have any connection whatsoever to the ALS mystery, or was it a missing piece to some completely unrelated puzzle?

And, let's face it. Pig scientists like Kristine may well know the answer. They may have written books describing it in exquisite chemical detail that I simply haven't stumbled across or been able to decipher. It would be fun to track down the research, discover the competing theories, and finally learn the answer. Why can pigs eat bracken? Why don't rattlesnakes kill the pigs that attack them?

That's what I wrote a few months ago. In a later chapter, I'll explain why pigs can eat bracken but I still don't know why they seem immune to rattlesnakes. As I was trying to learn about the remote and theoretical (from my perspective) relationship between pigs and snakes, reality forced my attention back into my own basement office.

Wild pigs. Photo by NASA, public domain

More Catfish Problems

My baby catfish started dying again a week or so after I cleaned their tank. The water looked clean and tested fine. Maybe some disease or parasite had infested it. With only two babies left and no good plan, I dropped an old penicillin tablet into the water and resigned myself to the inevitable. When the last two died, I'd have to start over with the tank, sterilize everything, and get some new fish.

It seemed like I was having a run of bad luck as a hobby biologist. One of the two toads my neighbor gave me seemed to have disappeared. Maybe it escaped but I didn't think so. Somehow I'd managed in a week to kill off a creature my neighbor had kept just fine for three years. A creature that can live 15 years or more. I did not mention the disappearance to my neighbor.

Even more embarrassing was the stench coming from my earthworm can. The soil remained damp, but the worms were all dead and rotting. Most ten-year-old fisherman can keep a few earthworms alive in a can. Clearly, I had lost my touch.

But the most bizarre thing was that the plants I'd dumped the nutrient rich aquarium water on were turning yellow.

I started to put the clues together. My aquarium water contained something toxic, either chemical or biological. I'm in the habit of being careful with fish tanks, although I'm the sloppiest person in the world in the rest of my life. Over fifty years, I've made most of the obvious mistakes and try not to repeat them. I could not imagine which mistake I'd made this time, but there was no doubt I'd done something wrong. If the water contained a toxin, it killed guppies and snails and catfish. Maybe the water I'd dumped from it into the terrarium had also killed the toad. Maybe the same toxin killed the earthworms. And maybe it was even attacking my garden plants.

You can kill tropical fish so many ways. A little soap on your hands when you feed them; Too much chlorine in the tap water you add to the tank. Parasites, worms, infections, temperature fluctuation, letting the pH get off, overfeeding, underfeeding, putting incompatible fish in the same tank. You do the best you can, but sometimes your fish just die. After a while, you stop agonizing over it and accept it as part of life. None of the possibilities would also kill earthworms, toads, and garden plants. Something else was going on.

The one bright spot was the baby catfish. Those last two did not die. Maybe I'd cleaned the tank well enough after all. Maybe they had overcome whatever disease or parasite killed all their siblings. Perhaps I'd solved that problem and could return to my book project.

Cyanobacteria were next on my list. If cyanobacteria in the roots of cycad trees create a toxin that can cause symptoms of ALS, certainly I ought to learn more about them. I'd heard of "blue-green algae" of course, but didn't know much about the stuff. Maybe they contained some clue to Suzanne's disease. As I began to read about them I learned they have a long and fascinating history, and modern scientists are spending hundreds of millions of dollars trying to harness their unique skills.

I confess, I did not become obsessed with tracking down the mystery playing out in my own basement. I was obsessed with cycad trees, snake-eating pigs, and the first life forms on Earth.

In retrospect, that seems ironic.

cory catfish
photo by Christian Ude

Cyanobacteria: the Early Years

Up until about three and a half billion years ago, Earth was pretty much identical to its uncountable siblings in the universe. Lifeless oceans churned and raged beneath its pink and yellow sky (although at sunset and sunrise the sky probably turned blue near the horizon). Lightning struck thousands of times a day, volcanoes belched fire and ash hundreds of miles into the air. Had there been anyone around to inhale it, the pungent air — thick with the gases ejected by volcanoes (like sulphur dioxide) —would have choked them within seconds. One gasp, perhaps two, and our imaginary time traveler would be dead. By mass, oxygen was the second most abundant element making up the planet, but it was nearly all bound up in molecules of water and other materials. The poisonous atmosphere contained almost no oxygen.

The lifeless earth conducted its furious experiments in weather, tides, and eruptions for millions of years without any audience that science can prove. Then, something miraculous happened. Perhaps God stirred the ocean with His finger; perhaps a hurricane combined chemicals like a gigantic food processor just as lightning struck and a volcano erupted. Whatever explanation you believe, you have plenty of company. But that's not the story I'm interested in right now.

Although we can't prove exactly how it happened, the biggest event since the dawn of time happened right here on our modest little planet. A very special kind of bacteria sprang into existence. In many ways, these little critters resembled every other kind of bacteria. Each individual was microscopically tiny– about two microns across (a micron is a millionth of a meter). If you stacked them on top of each other, you'd need a stack more than fifty bacteria tall to equal the thickness of a sheet of typing paper. Like all bacteria, they could reproduce at a fantastic rate: if nothing interferes, one bacteria can produce a million copies of itself in seven hours.

But these new bacteria performed three new tricks that changed the planet forever. They could process nitrogen, an inert gas that is completely useless to plants and animals, into forms that other life could utilize. Once the bacteria "fixed" nitrogen, combining it with hydrogen to create ammonia, then plants and animals could use the ammonia as a raw material to make proteins and other building blocks of all life on earth. To this day, they continue to fix nitrogen, sometimes in symbiosis with plants or coral.

These bacteria could also "fix" carbon dioxide. They could take in carbon dioxide and use it as a raw material. The chemicals that resulted would be useful to the bacteria themselves and also to other living things, once there were enough "other living things" on the planet to take advantage of it.

And they could use the power of sunlight to run the machinery of their bodies, releasing free oxygen into the atmosphere as a by-product, much as plants do today.

Millions of years later, algae and higher plants would arrive on the scene and duplicate the photosynthesis trick. Other bacteria would learn the nitrogen fixing trick. But back in the good old days of methane storms and red skies, these little critters changed everything.

First, they created the soup of organic chemicals from which the rest of life emerged. "Organic" chemicals aren't ones raised in a rural garden by a loving farmer. They are a large class of chemicals that always include carbon and often contain hydrogen, nitrogen, oxygen, sulfur, phosphorus, and other ingredients. Any substance created by a plant or animal is an organic chemical but some can also be created by lightning, heat, or other nonliving events. Without all those chemicals in the primordial stew, all future life forms would have had to learn to manufacture them. There's no guarantee they could have.

Second, these bacteria began to change the red skies to blue by churning out oxygen; as they did so, they cooled the planet. Oxygen diffuses a different portion of the light spectrum than the gases it displaced (which is why the sky changed colors) but it's not an effective "greenhouse gas" compared to methane, for example. The ancient pink atmosphere kept Earth nice and toasty, but the new oxygen-rich atmosphere created by these bacteria let heat radiate back out into space. For a while there, the sky was probably an intense blue as these bacteria released oxygen into a world where nothing breathed it. Many scientists believe this nearly froze the planet. Beneath the lovely blue sky, the oceans gradually froze and snow blanketed the land. This era is known as "Snowball Earth."

Ironically, the oxygen produced as an incidental metabolic by-product by these bacteria was poisonous to them. Humans produce chemical waste products that include carbon dioxide, carbon monoxide, hydrogen sulfide, and methane; in high concentrations, each one is deadly to us. Similarly, the oxygen that cyanobacteria produce can kill them. It's always a delicate balance.

Scientists believe that, during the Snowball Earth phase, huge populations of these bacteria died from exposure to the oxygen they produced. More died from the cold, others because the snow blocked the sunlight they required for photosynthesis. Only the toughest survived and they became even tougher. Always opportunistic, a few learned to breathe oxygen, transforming what had been a poison into a resource. Other creatures arrived that could also breathe the oxygen and reduce it from the toxic levels (for these bacteria) it had reached. The oxygen levels declined, the planet warmed, the determined little critters who lived through it began churning out oxygen and organic chemicals again.

We call these little creatures "cyanobacteria," although they were historically (and incorrectly) known as "blue-green algae." They made possible all the other life forms on earth. They provided the oxygen our ancestors breathed and the raw materials for plants and animals.

For their first few hundred million years on earth, cyanobacteria had it pretty easy. Living in mineral-rich oceans, they could produce food from the nitrogen dissolved in water and manufacture energy from sunlight. Best of all, they didn't have much competition. True plants, including real algae, had not yet developed. No animals, fish, or insects prowled the seas looking to eat them. Beyond churning out oxygen, their bodies also manufactured fats and oils, just as ours do. Not much, but some scientists now believe that over millions of years, those accumulated oils became the vast oceans of petroleum beneath the bleak sands of the Middle East. Some people are now experimenting with growing cyanobacteria to create oil to use as fuel.

Their microscopic lives became more complicated when little critters appeared in the ocean that could swim and eat. These opportunistic little hunters had stumbled onto one very basic survival strategy: "don't make it, take it." Why learn to fix nitrogen for yourself when you can just gobble up a juicy cyanobacteria that has already done it for you? Why learn to generate energy from sunlight when a tasty little cyanobacteria has stored up that energy in the form of fats, oils, carbohydrates, and protein? Especially when the oceans were teeming with them. Wouldn't it be more fun to spend your time inventing legs, fins, and wings? Without defenses, the cyanobacteria might have gone the way of the dodo bird and passenger pigeon: consumed into extinction.

But cyanobacteria weren't helpless. A few evolved the capacity to breathe this oxygen stuff themselves. Others produced chemical weaponry. They manufactured substances that tasted bad or that could irritate sensitive parts of their enemies. Some killed the predator outright. Creatures who made a living by eating cyanobacteria had chosen a very risky career path. Most animals chose other things to eat. Animals with a taste for cyanobacteria died of poison, leaving fewer and fewer descendants. Even today, only a few creatures intentionally eat them. Pelicans seem to thrive on them; so do a few insects. Some fish, like tilapia, filter any kind of algae into their system and don't seem to be hurt. Some higher animals ingest them with other food but don't seem to digest them; the cyanobacteria they eat pass through their bodies unharmed and largely unchanged. Other than that, unlike true algae, most aren't an important part of the traditional ocean food chain. They live, then they die, and other bacteria consume their little corpses. If they have made toxins, those are released into the water or air. Those toxins are remarkably stable. Boiling does not affect them. They may or may not have any taste or smell.

Not all cyanobacteria create toxins that are immediately dangerous to humans. Even those that can produce these toxins don't do so all the time. We have no idea why this is, or what triggers their urge to make poison. Even among individuals of the exact same species, some create toxins and others don't. You can't look at them under a microscope and predict which one will be poisonous. Only within the first years of the twenty first century did scientists start identifying genetic markers for toxin manufacture. It remains an infant science with more questions than answers.

True algae is, in many ways, a better menu choice. Each algae cell is much larger than a cyanobacteria, making it a more efficient snack. They are comparably nutritious. Best of all, true algae is rarely poisonous. You'd think the ocean creatures would have gobbled up all the true algae before it had time to proliferate. But algae has at least one big advantage over anaerobic cyanobacteria. Oxygen isn't dangerous to algae; it breathes the stuff. In fact, at night when it can't manufacture its own, it takes in oxygen from the water around it.

Imagine a battle for dominance playing out in a pond, featuring cyanobacteria and true algae. With everyone photosynthesizing, the oxygen levels could get high enough to kill the cyanobacteria, leaving the algae victorious. But then, if the pond contained lots of hungry bugs, the cyanobacteria might emerge the winner because its competitor tasted better and didn't kill those who ate it.

The earth has changed since cyanobacteria ruled it with a blue green fist. The air now contains about 20 percent oxygen. That's a dangerous level of pollution, from the cyanobacteria's perspective. The waters are no longer rich with nutrients the way they were back then. Other plants, animals, fungi, and bacteria compete for resources.

Yet cyanobacteria survive and even thrive. They live in the soil and the sea, in lakes and ponds and horse watering troughs. They form symbiotic relationships with plants and coral to fix nitrogen. In some cases, the toxins they produce protect those plants from predators. Cyanobacteria cling to survival in a world that no longer favors them. Despite all the oxygen and competition, they scrape out a dismal living in hard circumstances. Humans ignore them. Even congregated into huge globs, they are merely disgusting, smelly pond scum on the periphery of our world.

But don't feel sad for them yet. In one particular situation, nothing can compete with cyanobacteria. They like warm, still, water. Stir in a bunch of nutrients, like manure from a pig farm or runoff from an over fertilized field, and you've recreated the prehistoric waters they ruled for millions of years. Those conditions haven't existed on earth for over a billion years. No other modern life form evolved that long ago. All the other plants and animals developed much later, in a different kind of world. Cyanobacteria are tough little critters. They can clutch at a desperately poor environment and survive. But throw them back into the primordial soup of their ancestors and they reproduce with unmatched enthusiasm.

Man has become cyanobacteria's new best friend. We spread so much of their favorite chemicals onto our fields that our crops can't use it all. Rain washes the excess into ditches, creeks, and streams and ultimately into rivers and lakes. It has become so common to find bodies of water saturated with nutrients that we have a word for it. "Eutrophic" water is absolutely packed with nutrients, including phosphorus, which is fairly rare in a pristine mountain pond. We have inadvertently created the primordial soup of cyanobacterial dreams. Nothing on earth can out compete them in that environment. Like an alien life form from a distant galaxy, gasping for air and struggling against Earth's gravity, we have foolishly recreated their home planet, and then ignored them.

It's the opening scene from a bad horror movie sequel, only it's not fiction. In the first movie, perhaps called "Rise of the Cyanobacteria," these tiny creatures ruled an entire planet. In the second movie, perhaps "The Humans Strike Back," the empire of blue-green algae was beaten back into obscurity.

They've been waiting for three billion years. Last time, when the "higher animals" tried to eat them into oblivion, they were unprepared and vulnerable. This time they're ready. This time they're well armed. This time, the element of surprise is on their side. Could we be watching the opening credits of "Revenge of the Pond Scum?"

As I learned more, I began to feel like some ancient primitive blob of toxins stalked me from the shadows. How could I not know about this stuff? I imagined myself as a character in some grisly slasher movie, and the monster was visible to the audience, behind me in the mirror. As it turns out, it was closer than I ever dreamed.

Could My Tanks Be Infected?

Despite the common misnomer of "blue-green algae," cyanobacteria can be bright green, brown, or even red. They can float on the surface of a lake, pond, or ocean or they can stay below the surface. They can even live in dirt and the dry sands of deserts. They can float as individuals in the water, making it look like someone added green food coloring to it, or form long wispy chains. It's not surprising that they were originally considered "algae." Some varieties look just like it.

After reading descriptions of cyanobacteria for several days, I started to look at the stringy green algae in my catfish tank with new suspicion. I'd pulled out much of it and replaced some of the water several times, and the two baby catfish were doing fine. In all my years as a casual (and sometimes serious) tropical fish hobbyist I'd never, as far as I know, encountered "blue-green algae." Could it be possible, in the weirdest coincidence, that while I had been reading about cyanobacteria on my computer the stuff had invaded my aquarium a mere ten feet away? Could it be that brazen an enemy? My sarcastic inner voice said, "The stupid dinosaurs never had a chance." My sarcastic inner voice has never been much concerned with historical accuracy.

The algae in my tank was a bright green, not "blue green." It didn't float in the water like a fog of single cells; it coated the glass like moss and clung to the plants in wispy, web-like threads. Months earlier, one of my sons had given me two big snails to eat the algae. They cleaned half of it off the glass, then mysteriously died. Was that another clue I'd simply ignored?

48

I left the scientific websites and started exploring the tropical fish websites. Within thirty seconds it became clear that my entire tropical fish career had been lucky; thousands of hobbyists battled cyanobacteria. The news was grim. If your tank had cyanobacteria, most experts gave the same simple advice: give up. Remove your fish, dump your water, boil your sand, and start over. This stuff is tougher than you are, tougher than your fish, tougher than your drill sergeant in the Marines. Don't waste your time fighting it. Give up and start over.

If that wasn't an option for some reason, you had two choices. Because your opponent is not a plant but really a bacterial infection, an antibiotic might work. The preferred one seemed to be erythromycin. Perhaps, by sheer dumb luck, throwing that old penicillin into the tank had actually saved the catfish. Obviously, I had to do a little research on both penicillin and erythromycin. Their stories, believe it or not, would each make for exciting action/adventure movies and I'll tell you about them in a bit.

The other "cure" for cyanobacteria is oxygen. Remember that they evolved in a world without oxygen. Even though they may produce it as a by-product of photosynthesis, it's still deadly to them. Cyanobacteria prefer calm water; babbling brooks contain too much oxygen. Just stirring water increases the oxygen content, because more molecules of water get a chance to be on the surface in contact with air. Maybe this is why wastewater treatment plants often have fountains in their holding ponds. They need bacteria to decompose the waste, but they want to discourage cyanobacteria. An aquarium bubbler adds oxygen to the tank as well, but sometimes it can't keep up. To get rid of the infestation, you need to add a whole bunch more oxygen.

The easiest source of oxygen is hydrogen peroxide. It's cheap, relatively harmless to humans, and you can buy it at the grocery store.

A molecule of hydrogen peroxide is just a molecule of water with one extra atom of oxygen attached to it. Rather than H_2O (water), peroxide is H_2O_2. That extra atom of oxygen is not tightly bound to the molecule. With very little provocation, it will jump ship and roam around, ready to oxidize anything more interesting. This is bad news if oxygen is your personal kryptonite. Small amounts of hydrogen peroxide form naturally when sunlight shines on the surface of water.

Yes, hydrogen peroxide will kill cyanobacteria, but it will also kill guppies, catfish, snails, and water plants including regular algae. The trick is to

49

use enough to kill the bacteria but not so much it sterilizes your tank. The good news is that cyanobacteria are ten times as vulnerable to peroxide as algae or water plants are.

The bad news is that most of the websites discussing this seemed to be in England and they talked about micro liters and kilograms. I was not at all confident I could convert those into teaspoonfuls per ten gallon tank. And I didn't want to wait.

I drained about half the water from the tank and pulled out all the floating algae I could capture. I took a razor and cleaned the glass as well as I could. Then I took a paper towel soaked in hydrogen peroxide and used it to clean the glass above the water line. A small amount dripped into the water. After twenty minutes I refilled the tank with fresh water.

The next day, some of the water plants had white tips, as if I'd dipped them into bleach. But the glass stayed clean, the algae disappeared, and the baby catfish looked fine. The plants recovered. I noticed some very small snails busily cleaning the remains from the glass.

The parent catfish, now living in the new guppy tank, fell in love again. I managed to salvage 14 of their eggs. They hatched just fine in a smaller container and, when they were large enough, I put them into the tank that had killed so many. With no other change but cleaning out the algae, they have been thriving in it for the last several months. Dozens of snails crawl around the sand. Everyone looks content. I'm not sure what I'm going to do a few months from now when over a dozen catfish start falling in love with each other.

My opinion is this: I let the tank become eutrophic because I didn't clean the waste material out of the sand often enough. The water was saturated with nutrients. Then I inadvertently introduced some cyanobacteria into it with the duckweed from a pond. The bacteria thrived and released toxins that killed my guppies, then my catfish, then the snails. When I dumped that toxic water into my terrarium it killed the toad. When I dumped it on my can of earthworms it killed them. It even stunted the plants in my backyard that I dumped it on. But then I reduced the numbers of bacteria with the antibiotic and finally killed them off with hydrogen peroxide. If a few survived, they could no longer compete with the other plants because I'd removed so much of the nutrients. They no longer had a competitive advantage.

In retrospect, it seems so obvious. The most remarkable aspect is that all this happened to a guy who was actively studying cyanobacteria, keeping notes, and even writing about them. A guy who likes to think about science questions; a guy who was writing about solving science questions. And yet I didn't put the clues together until I'd lost a bunch of fish and spent hours worrying about it. It never occurred to me that the fish I fussed over in the morning had any connection at all to the book I was writing in the afternoon.

If I couldn't solve a simple problem that was right under my nose, when I had all the information and was actively studying the answer, it's no wonder that scientists haven't assembled all the clues about these diseases into one simple answer. There are too many clues and no one knows for sure which ones are connected.

I have new respect for the guys who do manage to solve these problems. They each focus, by necessity, on a particular part of the clues. And, like the blind men trying to describe an elephant by touching only the section of the beast they can reach, they tend to come up with much different visions of the problem itself.

The Possible Choices

Today, in the early years of the twenty-first century, no one can say for sure what causes a whole host of diseases that affect nerves: ALS, PMA, MS, autism, Alzheimer's Disease, and Parkinson's Disease.

Many scientists believe there is a genetic component. People might inherit one of these diseases, or at least inherit a predisposition to one of them. There is evidence this happens at least some of the time.

Some believe that a virus causes one, or all, of them. This idea originally arose from what seems like common sense. Some of these diseases act a lot like polio, which is caused by a virus. Just like polio, they destroy nerve cells which causes muscular atrophy, lack of control, or loss of some aspect of mental function. In the case of Guam, a virus spreading in a community might explain why it concentrated there. Similarly, that could explain the concentration among gulf war soldiers. They all caught the virus over there.

Some people believe these neurological diseases were caused by diet. The fact that the Chamorros of Guam ate some things that no one else did remained intriguing. But ALS strikes people around the world and most of us don't eat cycad nuts.

Some believe that something toxic in the environment is to blame. In an unscientific way, it seems like all these diseases are increasing. We hear about ALS and Alzheimer's more than we did when we were younger. But then, people live longer now than in the past. Maybe if people had lived longer in 1850 more literature of the time would mention it. Autism and Parkinson's seem to be increasing, but maybe our parents and grandparents called them something else. In Victorian times, women complained of "the vapors" and those seem to have vanished. On the other hand, migraines never appear in the literature of the day. Perhaps only the names have changed.

Still, it feels like these diseases have become more common and it seems like our environment is becoming increasingly polluted with toxins of all sorts. Therefore, instinctively, we suspect some industrial chemical.

Others believe a bacteria might be involved. Lyme Disease is caused by a bacteria (specifically a spirochete) that is carried by ticks. Once inside our bodies, the bacteria hides deep within tissue where antibiotics can't easily reach. It produces a toxin that causes muscle weakness, similar to these diseases. It's hard to diagnose and difficult to overcome.

Figuring out Lyme disease seems a lot like figuring out these others. Scientists had to first figure out what the victims had in common. As it turns out, they'd all been in the woods or owned a pet that had been in the woods. They had all been bitten by a tick that carried this bacteria. To solve the riddle, scientists needed to ask the right questions. Questions like "What did you eat for dinner the night before your symptoms appeared?" got them no closer to the truth. Even if everyone ate at the same restaurant and had the same dish, it had nothing to do with the disease. It didn't matter if they all liked polka music; that was sheer coincidence. When looking for the source of a disease, who would think to ask, "Did you pet a dog that had recently wandered through the woods?" Yet that was the common thread.

Just knowing the cause didn't necessarily mean you could cure Lyme Disease. You still had two tasks: killing the spirochetes and getting rid of the toxins they had dumped into your body.

Scientists still don't agree about these neural disorders. More than one thing might be the "cause," or several circumstances might join forces to cause these diseases. Recently, scientists have uncovered some startling new clues, as well as used some remarkable creative thought on the information we've already talked about.

We have a whole bunch of "maybe this is a factor" possibilities and a couple of "absolutely this can cause it at least some of the time" possibilities. Viruses, genetics and bacteria might cause the symptoms of ALS. They certainly might be contributing factors. But Spencer's experiment of feeding BMAA to monkeys proved that toxins absolutely can cause these symptoms in some situations. Yes, the dosages he used were impossibly high. Because of this, some people thought his experiment a failure. The natives of Guam couldn't get those dosages by eating cycad nut flour.

Looking at it a different way, the experiment was explosively successful. It proved beyond any doubt that the specific toxin BMAA could, in the right situation, induce the exact symptoms of ALS, the "lytico" part of lytico-bodig.

The Guamanians with "bodig" experience the symptoms of Parkinson's disease. A few people (10-15 percent) inherit Parkinson's. So we know there can be a genetic component. Depending on your perspective, that is either persuasive evidence that all cases are probably genetic, or else a misleading clue. If we could demonstrate that Parkinson's symptoms could be induced by some toxin on Guam, we might be able to feel optimistic that we're on the right track to understanding lytico-bodig and perhaps ALS and Parkinson's.

In 1976, an American graduate student's costly mistake proved that Parkinson's symptoms can be induced by a toxin.

Only four years after Woodstock, the hippie culture of American youth was beginning to morph into the disco era of good jobs and disposable income. Recreational drugs remained an acceptable part of many young people's lives. Barry Kidston, a 23 year old student at Maryland University, read a paper describing a synthetic type of opiate called MPPP and decided to mix some up and try it. He succeeded, but did not understand how important the temperature was at one stage and wound up brewing up a batch that had the impurity MPTP. He injected himself with his home brewed drug. He didn't realize his mistake until three days later, when he developed the symptoms of Parkinson's.

Luckily, he responded well to levodopa, a drug made from the velvet bean plant (*macuna pruriens*, a tropical legume that employs bacteria in its root system to fix nitrogen. Widely used as fodder for ruminants, such as cows, the plant is toxic to humans unless processed first by soaking). Levodopa, or L-dopa, is a precursor to dopamine, a chemical the brain makes. Unlike dopamine, it can cross the brain/blood barrier. With the help of L-dopa, Kidson lived for three more years without symptom until he died of a cocaine overdose.

Kidson's "experiment" was validated in 1984 when a group of scientists led by J. William Langston injected MPTP into squirrel monkeys. The monkey's all got the symptoms of Parkinson's. Interestingly, rodents did not. All the research that has sprung from this would require an additional book.

The chemical that Kidson accidentally brewed has become a standard tool in the quest to understand Parkinson's. Scientists administer it to laboratory animals; in every case, the animals develop the symptoms of Parkinson's. Then the scientists experiment on the animals to discover how different treatment strategies affect them. In some cases, they treat the animals first to see if they can prevent the disease.

Kidson's "experiment," like Spencer's, proved one thing: toxins absolutely can induce Parkinson's symptoms.

But the Guamanians probably weren't mixing chemicals in the jungles of Guam while hiding from their Japanese invaders. If a toxin caused "bodig" it wasn't MPTP.

Oregon Trip

I took a two week "vacation" from this little obsession to supervise my three grown sons as they mowed my father's weeds in Oregon. My father was spending time with my sister in Kansas, his "lawn" was six feet tall and going to seed. He owned over ten acres of "lawn." In Colorado, untended lawns simply die, quietly and with dignity. In Oregon, apparently, they go crazy. I was astounded. My father's land is by no means a monoculture of bluegrass. Dozens of varieties of grass, vines, and weeds competed for sunlight including many with thorns and stickers.

We made it a little father-son working vacation with my sons doing the work and me doing the vacationing. And buying the vast quantities of groceries.

One weed looked vaguely familiar to me. It reached above even the tallest grass, well over my head. It had pretty clumps of little white flowers that reminded me of the yampa (wild caraway) that grows in Colorado. Yampa is known as a traditional food of the Native Americans, and also as a medicine. I crushed a bit of a leaf between my fingers to smell it, but there was no familiar defining odor. I was pretty sure it wasn't yampa, because that smells like caraway seeds. I tossed it aside and made a mental note to ask my father about it. The stuff, whatever it is, seemed hardy in that climate. Dozens of the plants thrived on his property. It would be cool if it was useful in some way.

The morning we left, we got up at 4:30 to get an early start. Someone flipped on the TV to check the weather. The local station had a little human interest story about the weed I'd been wondering about. Yes, it's very hardy. Yes, it grows tall and develops pretty little white flowers. They showed lovely pictures of it. The name of the plant was "poison hemlock."

That's right. Hemlock, the deadly poison made popular by Socrates. Seven leaves will kill you and the roots and seeds are even deadlier. I was very glad I hadn't decided to brew a nice pot of tea for my family out of the stuff.

As we were driving away, my son Joey suggested that maybe the universe had drawn me to Oregon because hemlock was important for my book. Not one to taunt the universe, when I got home I did a little research on hemlock.

The roots are big and fleshy and the most poisonous part of the plant. Many accidental poisonings result from people mistaking hemlock roots for parsnips. Yes, the plant contains a powerful neurotoxin that paralyzes your muscles, beginning with your feet and legs and then progressing upward until it stops your breathing muscles. Interestingly, your body can actually process the poison if you give it enough time. If you get to a hospital in time and they put you on a ventilator to replace your breathing muscles for a few days, you've got a good shot of recovering completely.

Other than the fact that it's toxic and interesting, I found no connection between hemlock and anything I was studying. So I found myself randomly reading a series of pages about poisonous plants, fascinated in the way a twelve-year-old boy would be, not the way a scientist would be.

Scanning through the poisons on one such page, I noticed the word "legume" and stopped. Legumes have bacteria in their roots that fix nitrogen, some of those are cyanobacteria, so this seemed an interesting coincidence I should probably investigate. The plant was the "jicama." I'd seen jicama in grocery stores, a light colored round root bigger than a softball and popular in Mexico. I did not recall ever eating one.

The part of the jicama plant you buy in the store is apparently completely safe. In fact, as I discovered later, it's got some special qualities that make it very interesting as a healthy food. But all the rest of the plant is poisonous, especially the seeds. The poison is called "rotenone," which I'd heard of. I decided to check it out.

Rotenone is toxic to humans but it can't get through our intestinal walls very easily so we don't worry about it much. Fish and insects have different internal setups and it's deadly to them; it enters fish through their gills. It's been used in some form since at least 1849, primarily to kill fish or insects. The chemical itself was first isolated in 1895 by Emmanuel Geoffroy, and patented in 1912. Early in the twentieth century, a Japanese scientist named Roten studied this chemical, and it was named after him.

Rotenone has some advantages over other insecticides. Although it kills bugs slowly, it ruins their appetite right away; they stop munching on your garden plants. A few days of direct summer sunshine destroys it. It doesn't seem to affect humans much when we eat it. It binds to clay, so if you spill some on the dirt it won't get into the water supply or kill all your earthworms. By the 1920s, thousands of tons of rotenone were being produced. Then, we discovered more powerful insecticides, partly as a result of our military research during World War I on toxic gases. After World War II, we started to dump these new chlorinated insecticides (like DDT) on our fields and fruit trees. The relatively benign rotenone was forgotten.

In 1962, Rachel Carson published a book called "Silent Spring" that alerted the world to the dangers of chemicals like DDT in our environment. One of the book's readers was President John F. Kennedy, who instructed his scientists to check out her claims. Within ten years, DDT was banned in the United States.

Farmers still needed insecticides, just safer ones. Rotenone sprang back into widespread use. For years it was considered a miracle insecticide and even approved for organic farming.

Then, in 2000 scientists discovered something alarming. Large doses of rotenone cause Parkinson's in rats. Several scientists, including Betarbet and Sherer, conducted experiments that demonstrated this clearly, but exactly how it works or why remained unclear. We know that rotenone can pass through the protective blood/brain barrier. We suspect it might affect oxidative stress processes. But, as far as I can tell so far, we don't exactly understand what it does in our brains on a molecular level. We just know that big doses cause Parkinson's within some animals.

Because of this, in 2005 rotenone was removed from the list of insecticides approved for use by organic farmers. Once again, however, the common-sense logic that confounded Spencer's BMAA idea prevailed: no one was likely to ingest the dosages used in the experiment in the course of their lifetime. Therefore, it must be safe. Rotenone was once again quietly approved for use on tomatoes labeled "organic." It can be used on live chickens to kill mites. You might have eaten some today in your salad.

Then, in 2010, Francisco Pan-Montojo and Oleg Anichtchik did another study and decided that even tiny amounts of rotenone can induce some of the signs or stages of Parkinson's in "wild type" mice. I confess, I could not understand the abstract of this study completely beyond that, so I won't try to use it as "evidence" of anything. What it seemed to say was that, over time, the brains of all the mice given small doses of rotenone developed the characteristics of Parkinson's. Yet their bodies were able to process the rotenone so that none was detectable in their blood.

All this debate proved one simple fact: Perhaps several factors contribute, but we absolutely know that the symptoms of Parkinson's can be caused by toxins. And not just vague "environmental pollutants" released by evil multinational corporations in pursuit of world domination. One specific chemical created in a lab (MTPT) causes it every time. And one natural chemical, rotenone, does in large doses. This reminds us that one specific toxin, BMAA can cause the symptoms of ALS when given in large enough doses. This starts to feel like a pattern, or at least a clue. But it doesn't feel like an "answer" yet.

Yes, rotenone can produce the symptoms of "bodig" (which are also the symptoms of Parkinson's Disease) in very high doses. But the Chamorros hiding in the jungles of Guam during World War II weren't concocting fancy chemicals in their laboratories and they probably weren't doing a lot of gardening. They almost certainly weren't spreading questionable pesticides on acres of "organic" produce.

What they were doing was even more dangerous to their health, at least in my opinion. Here is one case where I noticed something from general literature and widely reported history that I have not found in any scientific studies about the disease. Let's see if you agree with me.

Derris Root

The Guamanians might not have been dusting tomato vines and arugula with rotenone while they hid from their Japanese invaders. But they were almost certainly fishing and probably using a cheap, effective traditional method.

The "derris" plant thrives in many parts of the world. For centuries native population (including the Chamorro) have used its root as a fishing aid. Fishermen crush the derris root, toss it into the water, then sit back and wait.

A chemical in the derris root intoxicates fish. They swim drunkenly to the surface, their reactions slowed, unable to swim quickly or evade the waiting fishermen who simply grab them with hands, net or spear. The chemical kills smaller fish, making them even easier to harvest. Short of using dynamite, it's probably the easiest way in the world to fish.

The chemical in the derris root is rotenone. I suspect Guamanians developed Parkinson-like symptoms at least partly because they sometimes ingested a whole lot of rotenone while fishing or by eating the drugged fish they caught. I have not heard anyone else suggest this.

Even though rotenone is considered safe for humans in small doses, this method of fishing has been outlawed in most of the world, probably for the same reason dynamite has been. Killing all the fish in a stretch of river just so you can eat one for breakfast is no longer considered responsible resource management.

We can't say that lytico-bodig is exactly the same as ALS and Parkinson's. But we can say that the symptoms seem to be identical.

We can't say that all cases of these diseases are caused by the same factors. But we can say this: certain toxins in high enough doses induce the symptoms of those diseases. BMAA can paralyze in the same way that ALS does. Either MTPT or rotenone can give someone the symptoms of Parkinson's. The Chamorros loved cycad nuts, which are a natural source of BMAA. And they used the derris root, a natural source of rotenone in fishing.

But rotenone is usually so harmless to humans it's approved as an insecticide on organic vegetables. Cycad flour doesn't have enough BMAA to induce symptoms.

If these toxins are to blame, we're still missing some pieces of the puzzle. The number of possibilities seems overwhelming:

One possibility: Maybe some people are genetically predisposed to these diseases. Maybe some people are naturally more sensitive.

Or perhaps we have a natural defense against the toxins. If someone happens to become exposed while this defense is weakened, a much smaller dose of toxin is required.

Or perhaps the toxins become much more dangerous when combined with some other factor. Perhaps a different toxin? I could not find any evidence that anyone tested BMAA in combination with cycasin, for example, but maybe they did. For that matter, I haven't seen any study of the effects of rotenone in combination with BMAA.

Or perhaps the toxins are much more effective when they enter the body through a cut in the skin, by a mosquito bite or by inhaling them.

Or perhaps there is some situation in which a person could get a massive dose of the chemicals without realizing it. Maybe the cycad nut is a clue, not the answer. Maybe the bucket where they crushed the derris root is the source of the contamination, not the fish they ate that night.

Maybe the the human body stores some toxins in a magical container that keeps us safe for years until it breaks and they are all released at once. Or maybe something produces similar toxins from within our bodies and that's why we can't isolate the source.

You see the problem. There are too many clues, too many choices. In the last dozen years, really smart scientists have proposed three or four radically different theories to explain all the facts. Each one seems completely logical and irresistible and each has an army of advocates. On the surface, they can't all be correct. The people who understand each of these ideas completely have spent half a lifetime or more becoming expert at some particular field of science; the problem is, they are all different fields. I can't imagine one person becoming a world class expert in each of these areas so they could reasonably compare them. You'd need to be an expert in chemistry, botany, the biology of the brain, the biology of the endocrine system, the biology of bacteria, toxicology, gastrointestinal disease, and nutrition.

You'd also need an unusually creative mind, one that was open to outlandish leaps. Luckily, several folks working in this field have exactly that sort of brain. One of them could be right.

Paul Alan Cox

In 2003, Paul Alan Cox, Ph.D., of the Institute for Ethnomedicine and his co-workers published a paper proposing a bizarre solution to the lytico-bodig puzzle that involved fruit bats. For his bold idea he attracted both positive and negative responses and the scientific community remains divided about how to react to his proposal. If he's on the right track, his idea may change the entire conversation forever. I've had pleasant e-mail exchanges with Dr. Cox and also with scientists who disagree emphatically with him. I've read their papers. I'll try to explain both sides of the argument. Luckily, you and I don't have to decide; we just want to understand.

Unlike the bats most of us know, the fruit bats of Guam don't use high pitched sonar (although they may use a lower pitched clicking sound). They have good eyesight and rarely eat bugs. They are primarily vegetarian; as their name suggests, they eat nectar and fruit. They can grow to be a bit more than one pound, which is a pretty good sized bat. It's about four times the weight of a brown bat, the most common in North America. The smallest bat in the world, the rare bumble bee bat of Thailand, weighs about as much as a dime. Fruit bats are sometimes referred to as "flying foxes."

Besides eating fruit, these bats also love to eat the juicy flesh surrounding the cycad seed. When they do (the theory goes) the BMAA toxin builds up within their bodies, reaching concentrations hundreds (and even thousands) of times that of the seed itself. This is called "biomagnification," the process of substances increasing in concentration as they move up the food chain.

We already know that mercury can become "biomagnified." Plankton absorb it from the ocean, plankton-eating fish get it from them. Contaminated fish get eaten by bigger fish who in turn get eaten. Each time, the mercury stays in the body of the predator after the rest of the meal is digested. In time, the mercury reaches dangerous concentrations in the fish's flesh.

Other toxins, like the ones produced by cyanobacteria, can become biomagnified as well. One such cyanobacterial toxin is saxitoxin. You wouldn't want to eat cyanobacteria (or the other tiny critters called "dinoflagellates" which are a bit more advanced) that produced it. But you really don't want to mess with it when it gets concentrated. Saxitoxin moves up the food chain until it reaches deadly concentrations in some shellfish.

Toxic shellfish poisoning kills people every year when they eat a few bites of the wrong seafood. Saxitoxin is a class one toxin - neither you nor I can make it or own it because its only use is as a deadly poison. During the Cold War, U.S. spies in high altitude U2 airplanes were equipped with a needle dipped into the stuff. If a spy fell into enemy hands and needed to commit suicide, that tiny needle would do the deadly trick.

The same magnification process occurs with fruit bats who eat cycad seeds, according to Cox. BMAA becomes concentrated in their bodies.

Historically, traditional Guamanians considered the fruit bat a great delicacy. They boiled them whole in coconut milk and ate them, fur, brains, meat and all. In one meal the Chamorros got a huge dose of the toxin BMAA. By analyzing tissue samples of fruit bats from the appropriate time period in a museum, they confirmed that the bats did seem to have a high concentration.

Cox created an entire theory to answer all the questions of Guam with fruit bats. Why wasn't the disease more prevalent in the decades before 1940? Perhaps, he suggested, before that the Chamorros hunting methods were primitive. Catching a big bat in a net on the end of a long stick is a tricky business. Hooking them on fishing line would be even harder. Delicacy or not, the natives probably only ate them on special occasions when the celebration was worthy of the hunting effort. During the Japanese occupation, guns certainly became more common on the island. Some probably fell into the natives' hands. After the war, as the natives became more prosperous, they bought guns. Suddenly, fruit bats were not such elusive prey.

Why did the disease disappear in the mid 1950s? Between over hunting and the tree snakes, the fruit bats began to disappear. The government declared them endangered and made it illegal to hunt them. After World War II , the natives adopted an increasingly western lifestyle. Teenagers weren't crazy about boiling up a big bat, fur and all, just because Grandma told them to. They'd rather have a burger and fries.

For the first time, we had a theory that explained most of the mysteries of lytico-bodig. Why it was on Guam but not Hawaii, why it became widespread, and why it disappeared. Perhaps most interestingly, it suggested that "biomagnification" could result in the huge doses of a toxin known to cause the symptoms.

But merely coming up with an idea that explains all the data is not the same thing as proving something scientifically. One could argue that a story in which aliens randomly injected natives with BMAA explains all the elements just as well. We can debate the relative plausibility of every scenario, but even plausibility is not proof. We needed more evidence, not just cool ideas. We needed reproducible experiments.

To help demonstrate the possible connection between BMAA and lytico-bodig, Dr. Cox and his team of scientists got permission to analyze the brains of Chamorros who had died of lytico-bodig. Each body contained BMAA.

A good scientist would argue that doesn't mean a thing by itself; maybe every corpse contains BMAA. For the study to mean anything, they needed to conduct identical examinations of a control group. The Canadian government agreed to provide some cadavers, none Chamorro, none from Guam, none who died from lytico-bodig. This should have been the boring part of the experiment. Either dead Canadians have BMAA in their brain tissue and we can throw out the theory, or they don't and we can proceed.

Instead, the plot thickened in an amazing and unpredicted way. Most of the Canadian cadavers showed absolutely no trace of BMAA. But eleven of them did.

All eleven died from Alzheimer's Disease. None of the others in the control group had. That is, among the Canadian corpses, every person who died from Alzheimer's had BMAA in their brain tissue. None who died from other causes did.

In a different experiment, Cox demonstrated that BMAA existed in the cyanobacteria in the roots of cycad trees. It did not exist in roots of the same tree without cyanobacteria. It existed in the cycad nut in much greater concentrations. The concentration was highest in the outer part of the nut. And the concentration was many times greater in the fruit bat. At each stage, as it moved from bacteria to nut to the animal that ate the nut, it became more concentrated.

Cox's idea has generated a lot of excitement. Many people believe he has solved the puzzle. Not only that, in his subsequent papers he may have answered why veterans of the first Gulf War seem to have the same symptoms as the Chamorro people.

Other people remain skeptical. I mean, really? Fruit bats? Some of the reluctance to embrace his idea has nothing to do with his science. Although he's got credentials coming out his ears, including a Ph.D., he's an "ethnobotanist" a scientist who studies traditional medicines and foods of native cultures seeking substances that modern medicine might want to study. It's a real science and he's a smart guy, but his specialty is not neurological diseases and he's making an argument about matters outside his field. Some bat experts disagreed that the fruit bats on Guam actually eat the cycad nut. Cox countered that they're right, the bats eat the soft outer part of the fruit- but that part also has the highest concentration of BMAA. They disagree with his estimates of bat population and about the number of bats that might have been consumed. Perhaps most troubling is that there are a couple of other isolated places near Guam that share lytico-bodig, but whose natives have never been known to eat fruit bats. They do, however, use unprocessed cycad nuts in folk medicine, including applying them to open wounds. Laura R. Snyder and Thomas E. Marker (Pathology Dept., University of Washington, Seattle) published a paper in 2011 saying that cycads produce BMAA even when raised in the absence of cyanobacteria. They feel that the attention given to BMAA is at the expense of all the other possible toxins, including others produced by the cycad, like cycasin, as well as other possible explanations for the disease.

In August, 2011, Glen E. Kisby and Peter S. Spencer, (the guy who demonstrated that BMAA can cause ALS symptoms in monkeys) published a paper in which they suggest that the BMAA in cycad nuts might work in a whole different manner. They suggest it might alter DNA and RNA in nerve cells and cause degeneration. Toxins that change DNA can cause tumors in cells that regenerate. Because nerve cells don't replicate themselves the way skin or liver cells do, we can't be sure of all the consequences a "mutation" like this might cause. They propose that this "genotoxic" action (instead of BMAA's activity as an "exitotoxin," killing nerve cells outright) could explain progressive neurological disorders and it might explain the slow development of lytico-bodig. Recently, they have become more focused on the other chemicals in the cycad nut, including cycasin and its by-product MAM, which also mess with a patient's DNA and RNA, especially if you're exposed to them when young. Cycad nuts have much greater concentrations of cycasin than they do BMAA. Because these chemicals cause mutations, they've been studied more in the context of cancer than brain diseases. Spencer and Kisby suggest that we should take a closer look at similar compounds in the typical western diet, like nitrosamines. Our bodies produce nitrosamines from some meat preservatives, like the ones in hot dogs.

Most who disagree with the fruit bat idea simply argue that Cox hasn't actually proved anything. It's an interesting idea, they say, worthy of more investigation, but it would be foolish to accept it as truth until definitive, reproducible experiments leave us no reasonable alternative. That's also the tone that Cox himself takes. Clearly he believes he's stumbled across something really interesting, but he's careful to limit how he interprets it.

After reading a whole bunch of articles by other smart sounding guys about all kinds of miracle cures for every disease in the world, it's probably a good thing that scientists maintain their skepticism until evidence forces them to release it. It's too easy for anyone, including guys like me, to find an audience for their odd ideas. If chewing the leaves of one plant cured diabetes and popping a cheap vitamin cured Alzheimer's, you'd think the medical and drug community would make sure we all heard about it. If not them, then the news media and the colleges. If no one else, the health insurance companies would surely give each of us a free supply of cheap vitamins just to save themselves the cost of treating all those diseases.

Cox's theory made a lot of sense to me so I decided to temporarily assume he was correct and see where that led me. Agonizing over this one theory didn't seem like a productive use of my time. I still felt like I needed more raw information. Just knowing about the fruit bats of Guam didn't seem likely to help my friend Suzanne any time soon.

Still, wouldn't it be weird if some guy found the key to ALS, Alzheimer's, and Parkinson's but nobody believed him because he had the wrong degree?

After I'd read several of Dr. Cox's papers and articles, I e-mailed him. He was completely gracious and kind, referring me to other resources, explaining things I didn't understand.

Most interesting was his response to a nagging question in my brain. The fruit bats eat the fruit surrounding the cycad nuts and develop high concentrations of BMAA within their own bodies. His theory is that BMAA causes lytico-bodig. That didn't make sense. So I wrote and asked him, "Why don't the fruit bats get sick?"

He e-mailed me back, "Good question."

Mosquito Fern

The last time I bought aquarium plants I'd gotten a few duckweed plant accidentally thrown in. Duckweed multiplies quickly; if they get enough light they will cover the surface of your aquarium before you know it. Some people find that annoying, but I don't mind at all. I like having the extra source of vitamin-rich food for my fish. But every week or two you need to clean out most of them or you won't have open water to drop food into.

That last batch of plants included some variety I'd never seen before. They looked like duckweed (which look like tiny lily pads a quarter of an inch across) but three or four times as large. They seemed to reproduce just as ferociously. I tried to identify them by searching the internet.

What I came up with is "mosquito fern." I'm not sure that's what was in my aquarium but that plant is fascinating and I'd never heard of it. They're called mosquito ferns because they blanket a body of water so densely that people thought they prevented mosquitoes from laying eggs. I haven't seen any scientific evidence that this is true.

Mosquito ferns (or "Azola") reproduce quickly to cover a pond because they have cyanobacteria called "Anabaena azollae" in their roots busily fixing nitrogen. In fact, across much of Asia farmers use mosquito ferns in their rice paddies the way American farmers use alfalfa or clover: to improve the soil. When the little plants die, they release this nitrogen into the rice paddy making it available to the young rice plants. Farmers buy bags of dried bacteria to use as starters for their rice paddies. It's a huge business.

The only disadvantage of this is that the mosquito ferns are toxic to livestock. The cyanobacteria in their roots produce a poison; I wondered if this poison was why farmers don't see mosquito larva in the paddies but I never learned if that's true. When the ferns die, they release the toxin into the water. You can't let your cattle drink that water. If they eat the plants or drink the water, they will become sick, paralyzed, or die. No one seems particularly concerned that the toxin might make its way into the rice itself, so it must decompose or bind with the soil in the bottom of the rice paddy. Or perhaps it shows up in such small traces that it seems harmless to humans.

It did seem curious that I'd stumbled across another plant with cyano-bacteria in its roots fixing nitrogen that also produces a toxin that paralyzes animals. Weird that it would be produced in vast quantities around the world and important to agriculture.

And odd that some should arrive at this particular moment in my own troubled aquarium.

Microscopic Paradoxes

At some point, when you're gathering information in a haphazard way like this, your mind boggles. At least mine did. You want to collect information, but you have no way to know which facts are important, let alone which are connected. You want a road map, like you had back in school. Back then, smart people had already assembled the facts into neat compartments: "Medieval Literature," "The History of the Roman Empire;" and "Impressionist Painters in France in the 1890s." Much harder to find patterns when you read a few paragraphs about each in quick succession. Scientists avoid saying they "know" something until they can prove it with an experiment that others can replicate. The rest of us just want to "know."

I felt like I needed to organize my confusion by identifying it. Some of my mental sea sickness came from a rolling tide of paradoxes. Maybe it would help if I simply listed them.

The world of microscopic life confounds the mind with opposites. For example, each individual critter is too small to see with the human eye but, if you collected them all in one place, they'd be an unimaginably large monster. One article said that the aggregate weight of all the bacteria on earth is greater than the weight of all the plants and animals combined. I'm not sure I believe it. Still, even if that's not accurate, we can all agree that the number of microscopic critters on earth is "a lot."

Some bacteria cause terrible diseases. Others manufacture deadly toxins.

Yet bacteria also digest our food, manufacture vitamins, transform milk into cheese and grapes into wine. A factory line of bacteria within cow stomachs digest hay. They process sewage and decompose dead animals, returning their raw materials into the cycle of life. Some detoxify poisons by eating them. They fix nitrogen and carbon dioxide; they create oxygen.

Approximately 2,000 varieties of bacteria live within the human body. Of those, only about a hundred are "enemies." The other 1,900 or so either help us or are merely along for a free ride.

Bacteria within our intestines provide most of our immune systems. We can't survive without the "good ones," yet the "bad ones" routinely kill us.

Bacteria may well be the oldest form of life on our planet. And yet the study of them is one of the newer sciences. The other sciences are hundreds if not thousands of years ahead of the study of bacteria.

The Egyptians used the principles of physics six thousand years ago. The Mayans and Babylonians studied astronomy. The Chinese employed primitive chemistry to develop fireworks a thousand years ago. The Greeks and Romans studied plants and animals. Da Vinci studied anatomy in the 1500s.

The first human to glimpse and describe bacteria was Antony van Leeuwenhoek in 1683. But it wasn't until nearly two hundred years later, in the 1850s than someone came up with the idea that these little critters might cause disease. Until then, they remained scientific curiosities. Even after that, our knowledge remained fairly primitive. In the late 1800s we came up with some good ways to kill them, like using iodine. The first antibiotic wasn't discovered until the 1940s. Humans have been studying chemistry, physics, astronomy, and the biology of plants and animals for well over a thousand years. In comparison, we learned nearly everything we know about bacteria in the last hundred years.

Another paradox is how much we know about some bacteria and how little we know about others. A few bacteria have been studied intensely. We know the chemical processes of their tiny bodies; we know the sequence of genes in their chromosomes. We know where they thrive and how to kill them.

Most bacteria remain much more mysterious. Of the 100,000 or so varieties (there is some controversy about whether using the term "species" is appropriate for bacteria) that have been identified, only a handful have been studied in much depth. The figure "100,000" refers to the varieties that we've identified at all. Many scientists estimate that 1.5 million varieties exist. If they're correct, we haven't even named more than 90 percent of them, let alone studied them.

Take the cyanobacteria. We've studied a few dozen of the 4,000 varieties believed to exist. The average cubic yard of dirt contains 400 kinds of cyanobacteria. Some varieties live in salt water, still others in fresh water.

Not all varieties of cyanobacteria seem to make toxins. And the bad ones aren't predictable: sometimes they're dangerous and sometimes they aren't. No one yet knows why this is. We don't know what triggers the production of poisons, we don't know how to prevent it. You can't look at a cyanobacteria under a microscope and determine if it's poisonous. We are just beginning to identify genetic markers for toxin production.

In fairness (and as an example of another paradox), I must give a nod to the many sincere people who consider blue-green algae a health food. Of the thousands of varieties of cyanobacteria, a specific one, "spirulina," has been used as food for centuries. The Aztecs ate it, as did people in parts of Africa. Spirulina leaped to prominence in the mid 1970s, when people at the United Nations World Food Conference suggested it as a promising food crop in a world of diminishing resources.

Then, in the 1980s, some folks started promoting it as a fantastic health food, full of nutrients and near-magical health benefits. They sold a lot of spirulina. But courts ruled that their claims were unsupported, they had to pay huge fines. Other companies sprang into business. Some of the new ones seemed suspiciously related to the earlier enterprises that had been scolded by the courts. I don't want to discourage anyone from buying a product I don't know anything about and I haven't read much about this whole industry. It looks like pure spirulina does not produce these toxins. When grown in a controlled environment, I don't have any reason to suspect it might be dangerous. If other varieties of cyanobacteria might be mixed in with it, I'd be less confident. If a scientist can't look at the stuff under a microscope and determine whether it will produce toxins, then I don't feel confident I can tell just by looking at the box it comes in. Yet many people swear by this stuff.

Paul Cox and his team tested many kinds of cyanobacteria to see which ones produce BMAA, the toxin they suggest might be responsible for ALS on Guam. Some varieties produced a lot, some just a little. But over 90 percent of the thirty varieties they tested produced at least some BMAA. As far as I can tell, spirulina was not one of the kinds tested by Cox, but an earlier test showed that it did not manufacture the major toxins that have worried people for decades. I could find no information indicating whether spirulina's been tested for BMAA or not. As to the pond scum at your local park, the odds are not so good. It is almost certainly not pure spirulina. There's a pretty good chance the green muck in your back yard is manufacturing various toxins, including BMAA, to at least some extent.

In response to which I hear you thinking, "so what?" You don't eat a lot of blue-green algae, and you only drink water that your local health department has processed and declared safe. If you're a bit more paranoid, you only drink bottled water because that is probably even safer, being in a bottle and all. You don't swim in muddy ponds and you only eat organic vegetables. Sure, one kind of cyanobacteria that lives in dirt and produces BMAA ("Nostoc flagelliforme") is actually eaten as a delicacy on Chinese New Years. It looks like an appetizing wad of black hair and is called "Fat Choy." Eaten just once a year, it doesn't seem to cause problems, but now it has been outlawed. You don't care, do you? You weren't in China celebrating that particular holiday. Even if you were, you probably avoided most of the dishes that looked like wads of black hair.

To most of us, this is mildly interesting in a "Gee, I'm glad I don't live in some Third World country where they drink dirty water" kind of way.

Our smug sense of safety may be premature.

Pond Scum

If you stroll past any pond or lake in North America on a hot day in August, you may see a mat of bright green algae covering some part of the surface. This might be a "bloom" of harmless algae, or it might be a "bloom" of cyanobacteria. Your local newspaper may issue a warning to keep your children and pets out of this pond, especially if it's in a favorite local park, although they will probably refer to it by the older, friendlier-sounding term "blue-green algae." Every mayor and Chamber of Commerce would rather acknowledge a lovely bloom of algae in the local park (along with the rose blooms and daisies) instead of advertising a seething mass of toxic prehistoric bacteria. The weekly newspaper will caution that algal blooms can be harmful to pets and cause skin irritation in humans. A week or two later, the bloom will be gone, officials will test the water and declare it safe. Life will return to normal.

These warnings are a good thing. Every year, pets and livestock die because they drink water poisoned by cyanobacteria. Recently, six dogs died in the state of Washington after swimming in a lake and then merely licking their fur. The history of unexplained livestock deaths is littered with victims and suspected victims. These deaths are caused by the dramatic fast-acting toxins these bacteria churn out. The toxin BMAA that has been implicated in both ALS and Alzheimer's — the toxin found in cycad nuts and produced in the plant's roots by cyanobacteria — is not one of those fast-acting poisons. It seems to build up within tissue over time. Human deaths and sicknesses are on the casualty list too, but not as many as you might guess. Humans avoid water that tastes bad, and pond scum tends toward the unsavory.

One of the chemicals cyanobacteria produce, geosmin, gives pond scum its foul taste. Interestingly enough, this is also the chemical that produces that fabulous organic smell when rain finally hits dry earth after a rainless period. That smell has its own name, "petrichor." Geosmin gives us the rich organic smell of good soil in our garden, the petrichor, and humans are extremely sensitive to the smell. Some people can detect it in air diluted to as little as five parts per trillion. But in water, we hate it. Our sensitivity to this smell and taste probably helped humans survive millions of years of attempted murder by cyanobacteria.

But that doesn't mean we're safe. To understand why, we need to go into a bit more detail about the complex drama that unfolds in the pond a few blocks from your home.

It's a beautiful August day; the sun blazes above us. We're in your favorite park strolling past your favorite duck pond. The maintenance folks have been fertilizing the grass more than usual, the excess has washed into the pond. The water is saturated with nutrients, which is called being "eutrophic." Underwater plants and algae love fertilizer as much as lawns do; so do floating plants like water lilies and duck weed. The pond is green as a golf course.

All these plants use sunlight to fuel their growth, all produce oxygen. If we look closely, we can see tiny bubbles of oxygen floating to the surface and exploding into the air. Plants, including algae, need to "inhale" that oxygen at night. So do all the fish, crawfish, snails, and worms in the water. Luckily, the plants create more oxygen during the day than they all need to survive the night.

But, because of the fertilizer, the plants have grown enthusiastically and have nearly filled up the pond. There isn't enough water to absorb all the oxygen they create, the excess bubbles away. At some point, there are too many plants and animals trying to breathe in too little oxygen every night and they all suffocate. A few cloudy days might speed this up. Now you've got a pond full of dead, rotting plants. The bodies of dead fish float to the surface, adding a festive touch to the muck and stench.

The bacteria that consume the carcasses don't produce oxygen. Some use oxygen, some don't need it, but none produce it. Over fertilization caused the problem, but the cause of death was suffocation. It's a stinky mess, but it probably won't kill your dog if he decides to take a swim. Water without oxygen is called "anoxic." Ironically, plants that produce oxygen led to the anoxic condition of your pond by blooming then dying, and it won't recover until the dead plants decompose and new ones replace them.

A cyanobacterial bloom is very similar. In fact, it may look exactly the same to you or me as we wander past the pond. But there are a few key differences.

First, some cyanobacteria contain little air sacs which they can regulate. By inflating these, they control how deep below the surface they float. This ensures they get the optimum amount of sunlight. You might not notice them as they float two or three feet below the surface. Then, one day, they appear as if by magic on the surface and you are astonished that they came from nowhere.

Second, they don't need to breathe oxygen at night. Oxygen is toxic to most cyanobacteria. The bloom is limited not by insufficient oxygen, but by too much oxygen. Luckily for them, when they decide to float on the surface, most of the oxygen they produce is released into the air, not into the water. A thick mat of cyanobacteria shades any plants beneath the surface, preventing those from producing much oxygen either.

Third, the cyanobacteria may produce toxins. Their toxins fall into three categories: neurotoxins affect the nervous system, hepatoxins affect the liver, and dermatoxins affect the skin. The dermatoxins cause "swimmers itch."

72

Plants, including algae, have cell walls composed of cellulose. Fungi have cell walls made of chitin, which is a lot like the stuff in your fingernails or the exoskeleton of an insect. Cyanobacteria have tough cell walls made of protein. This wall protects the complex enzyme they use in fixing nitrogen from oxygen, which would otherwise destroy it. The cell wall helps contain the toxins created by the bacteria. Most of the toxins remain within the bacteria until it dies and the cell decomposes, a process called "lysis."

That is, most toxins aren't in the water during a bloom. Most are in the water when the bloom is dying and the cells are breaking down. The toxins can stay in the water for three weeks or more (longer if it's not sunny) until different bacteria have their own little invisible bloom while they consume them.

Both algae blooms and cyanobacteria blooms create "dead zones:" areas where nothing can survive. An algal dead zone will recover when plants reappear to replenish the oxygen in the water. A cyanobacterial bloom involving toxins can recover only after a different bacteria digests the toxins and then plants replenish the oxygen. For most purposes, people (and media reporters) don't distinguish between the two. The water smells bad, nothing lives in it, and it's depleted of oxygen. But if your dog drinks water from a lake with a cyanobacterial bloom, he may die. That's the big difference.

Every summer, scientists track about 400 huge "dead zones" around the world. In the public reports, they usually don't distinguish between algal blooms and cyanobacterial blooms. The bloom in the Gulf of Mexico, fertilized by the Mississippi River often exceeds 8,000 square miles. That's about the size of New Jersey. One in the Baltic Sea is about the size of Germany. The Red Sea gets its name from the red cyanobacteria that live in it. When they bloom, it's called "The Red Death."

The lake that provides drinking water for your city isn't immune. In 2010, cyanobacteria bloomed in Indiana reservoirs that serve a million households. Also in 2010, a bloom killed nearly all the fish in the St. John River in Jacksonville, Florida. In 2011, a bloom in Lake Calavaras stunk up the water for citizens of San Francisco. Virtually every state issues warnings about a few "toxic blue-green algae blooms" every year. Many of these occur in reservoirs upstream from someone's kitchen faucet. In 2011, Lake Erie had a huge cyanobacterial bloom. The layer of scum was four inches thick and covered 2,000 square miles. Eleven million people get their drinking water from Lake Erie.

The good news is that scientists have known about cyanobacteria toxins for years. Modern water systems have fancy charcoal filters that can remove the toxins.

The bad news is that using these filters is expensive so water departments only use them when they discover one of the toxins the EPA requires them to test for.

The further bad news is that cyanobacteria can produce as many as 80 different kinds of toxins. The EPA only requires your water department to test for three of these. BMAA isn't one of the three.

The worse news is that these toxins are colorless, tasteless, and odorless. Boiling does not neutralize them at all. It may actually intensify them, perhaps by boiling away some water and leaving the remaining toxins more concentrated. We know that some persist in our bodies for years. Perhaps they build up within us, increasing in concentration just like they do in fruit bats. We know some of them are quite nasty, even deadly.

Yet I'll bet 95 percent of American adults haven't spent two minutes of their lives thinking about cyanobacteria or their poisons.

Honeybees

Like most writers, I practice being creative, stretching the mental muscles that notice weird connections, unique metaphors, and goofy ideas. That does not, by itself, make me crazy. I relay the following thought that occurred to me with that in mind. It's probably nothing.

The honey bee population has been decimated by a mysterious illness called "colony collapse disorder." For the last few years, entire hives of bees simply disappear. Because bees pollinate so many of our crops, this is a multi-billion-dollar concern. Scientist thought mites could be responsible, and that's all we heard about for a while. Then they thought it was a virus. Now some suspect that a combination of a virus and a fungus is the culprit; others blame a parasitic fly. People who maintain a few hives as a hobby are experimenting with that idea by breeding bees that instinctively "clean up" infected pupae, which interrupts the fly's life cycle. They see some encouraging results. Others suspect that some sort of poison is to blame, and they focus on one kind of poison in particular.

"Neonicotinoids" are a type of insecticide widely used around the world that keep coming up as possible culprits. They affect the nervous system of many kinds of animals, but insects more than others. They are used on grains, vegetables, potatoes, cotton, fruit, and grass. Some forms can stay active for years in the right conditions. They can spread through a plant to its leaves, flowers, fruit, and even pollen. Some people suggest that the poison from treated corn can even survive after it's processed into high fructose corn syrup. They point out, ominously, that commercial bee keepers often supplement their hives' diets with high fructose corn syrup. Could that be the source of the problem? It's apparently difficult to test high fructose corn syrup for these insecticides.

In one interesting application of these poisons, the insecticide is stuck to a seed (like corn) with a sticky goop before it's planted. Then the goop is coated with fine powder (like talc) so the seeds don't stick to each other or clog up the planting machine. Some people think this powder coating can get knocked loose by the planting machine, float into the air, and accidentally carry insecticide to pollinators. Even if they don't get enough to kill them, it may be enough to confuse and disorient the bees so they can't conduct normal bee business.

Because of this suspicion, neonicotinoids have been banned on and off since the 1990s in some European countries. Two or three books have been written about this theory; at least one documentary was filmed. The companies that manufacture these poisons insist they're safe. The bees continue to die and no one's completely sure why. An odd idea occurred to me.

It turns out that bees drink a lot of water, especially during hot weather. They love to drink from still ponds. The exact kind of ponds favored by cyanobacteria. Maybe sometimes they drink water polluted by toxins manufactured by cyanobacteria.

I wonder: could natural poisons that kill dogs and sheep harm honeybees? Could toxins like BMAA, made by blue-green algae, amplify other factors like insecticides, mites, virus or fungi? I don't know, I merely pass it on as an interesting idea. I haven't heard anyone else suggest this idea.

Resourceful Bacteria

You don't survive for three billion years without being resourceful. Cyanobacteria are the masters of resourcefulness. They spin life out of chemical muck and sunlight. They survive in steaming vents of poisonous water on the ocean's floor and sleep through the winters of Antarctica. When they sense their surroundings becoming uncomfortable, some of them create a stripped down version of themselves called an endospore (they are also called "akinetes") to lie dormant – for centuries if necessary – until conditions improve.

Being able to remain dormant during hard times is a huge advantage. When the next Ice Age comes along, cyanobacteria will simply nap through it. Ten thousand years later, when other species have become extinct, they will awake, refreshed, and cheerfully start reproducing like crazy.

An endospore resembles a virus in some ways. Either one can be described as a strand of genetic material inside a tough protein coat. When either one lands in a friendly environment, it begins to rebuild the cell that created it. But, unlike an endospore, a virus requires a host. When a virus lands in a friendly environment, it injects its DNA into the most convenient cell and hijacks it. The virus DNA directs the host cell's machinery to build new viruses. At some point the host cell bursts, releasing hordes of new viruses.

When an endospore lands in your nice warm garden pond, it wakes up and builds one new cyanobacteria. Given some sunlight and nutrients, that one critter will begin to reproduce without any other assistance. It could travel across space and be perfectly happy as the only life in a warm, barren ocean.

Some viruses manage to insert themselves, not just into a cell, but into the host cell's DNA altering it forever. This might be disastrous for the host. HIV virus works this way; it changes the human's DNA in ways that are very unhealthy for the human. Other times, this strategy could be disastrous for the virus. The altered gene may quickly result in a dead host and, indirectly, a dead virus.

Sometimes, merging DNA may have no important effect at all. Humans, like all creatures, have a whole bunch of genetic material that doesn't seem to have any purpose. Some of this may represent genetic history; your kitty may contain all the information required to recreate a saber toothed tiger, if only we knew how to turn the right switches on and off. That's the premise of the fascinating book, "How to Build a Dinosaur" by James Gorman.

Sometimes, we think, these invasions result in good things. It is possible (though not proven) that a photosynthesizing bacteria invaded a plant cell at some point, the two managed to reach a genetic agreement, and the bacteria became entrenched as a little organ within subsequent generations of plant cells. Its job was to convert sunlight into energy. These little organs are called chloroplasts. Some scientists think a similar scenario resulted in the little organelles within animal cells called mitochondria. Their job is to create energy for the cell.

Cyanobacteria have other survival tricks beyond cellular piracy and creating endospores. If you try to poison them, some will simply gobble up the poison and ask for seconds. In 2008, Ronald Oremland of the U.S. Geological Survey discovered that cyanobacteria were eating arsenic at the bottom of Mono Lake in California. A few billion years ago, he believes, arsenic was a pleasant snack for these fellows. You're not going to scare them that easily. Some of them can manufacture a protective shield called "biofilm." This little tent protects them from just about everything. There is evidence they can detect tiny amounts of chemicals in their environment, almost as if they smell them. The way they react suggests, to the imaginative mind, that individual cyanobacteria cells can communicate between themselves in a primitive way.

In 2008, scientists attached some cyanobacteria to the exterior of the International Space Station just to see if they could survive in outer space. In 2010 they retrieved them. The bacteria had been riding along up there for a year and a half, unprotected by atmosphere, subject to radiation and blazing heat alternating with near absolute zero cold. It didn't even dampen their mood. These particular bacteria were originally collected from cliffs near the English town of Beer. My sarcastic inner voice suspects these specific bacteria were gathered from that town for the benefit of Internet search engines. "Beer bacteria in space" will get more hits that any combination of "cyanobacteria" and "International Space Station."

Cyanobacteria live in the sand of the Middle East. They simply lie dormant during the dry times in the form of a mat that helps hold the sand together. When it finally rains, they spring to life and photosynthesize as fast as they can to store up food. When the sand dries out, they just go back to sleep. Compared with outer space or Antarctica, Kuwait is a five star hotel for them.

Veterans returning from the Gulf War displayed symptoms of ALS much more frequently than the general population. The soldiers certainly hadn't been eating the sand. But there had been terrible sand storms. Could the cyanobacterial toxins in the sand affect a person if he inhaled them?

Paul Cox, the guy who came up with the fruit bat theory of lytico-bodig, collected samples of cyanobacteria from Middle Eastern sands, cultured them in his lab, and tested them for BMAA. They did, in fact, contain the toxin. Cox wondered if inhaling BMAA (rather than eating it or injecting it) could cause the symptoms of ALS. We know that inhaling some toxins is far more dangerous than eating them. In some cases, inhaled toxins are between ten and fifty times more dangerous. During a violent sand storm, soldiers in the Gulf breathed in dirt and cyanobacteria and probably toxins. Dr. Cox suggests that inhaling BMAA may explain why they got sick.

Several neurological diseases are associated with a reduced capacity to smell things. Among these are Alzheimer's, Huntington's Chorea, and Parkinson's Disease. Reduced olfactory capacity is also a side effect of inhaling cocaine, a drug which interferes with the normal processes that recycle the brain chemical dopamine.

I thought of the Chamorro people on Guam, crushing derris root for fishing and perhaps sniffing it now and then to make sure it smelled right. And then cutting up cycad nuts, processing them into flour while hiding in the jungle without all their traditional tools. Cooking a fruit bat for their starving family. I can imagine many situations in which someone might inhale deeply to savor a scent without considering they might be getting a dose of toxin. Just like I had done with the hemlock.

Why Don't the Fruit Bats Get Sick?

Think about it: the fruit bats of Guam eat the cycad nuts, the BMAA becomes highly concentrated in their bodies. Why don't they all get sick? If that toxin causes all these neurological problems you would think it would kill them all. For that matter, why didn't the rats Spencer fed BMAA get sick? Without a good explanation for why fruit bats and rats seem immune, I won't ever feel completely comfortable with Dr. Cox's theory. So I decided to look for possible answers to the question.

One possibility is that it certainly would kill them if they lived long enough. Chamorros often developed symptoms twenty years after they left Guam. Perhaps BMAA is such a slow acting poison the bats die from other causes, including old age, before it can affect them.

That makes a lot of sense and it may well be correct, but it wouldn't do Suzanne much good. On the other hand, maybe something else in the lifestyle, diet, or genetics of fruit bats protects them. If so, maybe it would protect humans as well. I started trying to learn about fruit bats. It turns out, not much fruit bat science is written in language I can understand. But I gathered a few tidbits.

Fruit bats eat all kinds of fruits and nectars, but their favorite food seems to be breadfruit. OK, that was a start. I've never eaten it, but it sounded like an interesting fruit. It's a tree in the mulberry family that can grow to sixty feet tall. The fruit, a little larger than a cantaloupe, is consumed in tropical regions all over the world. Supposedly, it actually tastes like bread when you cook it. It was made famous by the book Mutiny on the Bounty. Captain Cook had seen it on Tahiti in 1769. Eighteen years later, Captain Bligh was sent there to get young trees to transplant to the West Indies as food for the slaves. But, in the six months or so his crew was collecting breadfruit saplings, they fell in love with the native girls and didn't want to leave.

The rest is history (and a book, and a movie). Ultimately, a later expedition succeeded at the mission and transported a bunch of breadfruit trees to the Caribbean. Unfortunately, the slaves didn't care much for breadfruit. I don't believe they made a movie about that. I added "breadfruit" to my list of things I'd like to sample some day.

The nutritional content of breadfruit is about what you'd expect: fiber, vitamin C, potassium, and B vitamins.

A deficiency in some of the B vitamins causes symptoms nearly identical to Alzheimer's. That might be worth investigating later, I thought. Vitamin C has been in the news recently. Some claim it's vital for cardiac patients, but it's controversial. I started by trying to learn about it.

Most animals don't need vitamin C in their diet. They manufacture it themselves in a process that involves four different enzymes. But we've known for a long time that four groups of animals lack one of these enzymes. If they don't eat enough vitamin C, they get a disease called "scurvy." Those four mammals are humans, gorillas, guinea pigs, and fruit bats. It's a common belief (at least on the Internet) that these are also the only mammals that suffer from heart attacks. We now know some other bats, primates, and a few birds also lack the ability to make Vitamin C. It's obvious how a vitamin deficiency became interesting to cardiac scientists. If you believed that the only mammals susceptible to heart attacks and strokes are also the only ones that can't produce vitamin C, you'd start to look for a connection.

Linus Pauling became famous partly for his enthusiastic promotion of Vitamin C. His many books and papers encouraged people to consider that the tiny amount of any micro nutrient needed to prevent some terrible disease might not be nearly the optimal amount. Most of the research down this interesting path relates to blood vessel health. Vitamin C is a powerful antioxidant; that may be why it seems to protect against heart disease. It indirectly affects nitric oxide levels, which affects blood pressure.

One group of researchers believes that a lack of vitamin C is the largest cause of heart disease. They claim that C is the ideal first responder to vascular irritation and only when that's lacking do our bodies send cholesterol to smooth over the site, ultimately clogging the artery. Are they crazy? I can't tell. I will say this: other mammals make a whole lot more vitamin C than humans eat. Goats, for example, naturally produce about 200 times as much vitamin C as humans have on a good day.

Luckily, the human body developed tricks to recycle it. We don't need goat-like levels of vitamin C to survive, but some people argue that we'd be much healthier if we got a much more than the minimum. The fact that fruit bats and humans need vitamin C seemed interesting, but I never found anything beyond that.

Gorillas, guinea pigs, and fruit bats are almost exclusively vegetarian. Maybe some interesting clues lurk in the shadows of a vegetarian diet. But for right now, I'm thinking about breadfruit.

Breadfruit contains a lot of choline. This was interesting because choline is involved, at least indirectly, with ALS. Our bodies use choline to create acetylcholine, a chemical nerves use to communicate with each other. The few drugs approved for PMA and ALS slow the destruction of acetylcholine. An ingredient in green tea does something similar.

For decades, choline has been suggested as useful in slowing Alzheimer's Disease. Patients with Alzheimer's have reduced levels of acetylcholine; perhaps eating choline gives them raw materials to make more. There's some evidence that it's useful, but obviously not a "cure." The fact that fruit bats eat breadfruit, which contains choline, which is involved with nerve chemicals meant I had to read up on this stuff. The easiest way to get extra choline into your diet is to eat lecithin.

Lecithin is a component of cholesterol. It's found in egg yolks, soy, and many other foods. If you buy it at the health food store, it resembles tasteless honey; thick, sticky, gooey stuff that isn't especially pleasant to eat. You can also buy it in granular form to sprinkle on your food. Lecithin has become popular recently because (according to many seemingly reliable folks) it can dissolve the cholesterol that builds up in our arteries, ultimately transporting it to our intestines where our bodies eliminate it. Some people claim that taking two tablespoons of lecithin daily for two weeks can actually dissolve the plaque buildup within arteries. I've added lecithin to many of my recipes, just in case.

The idea that lecithin can dissolve cholesterol and then cling to it in a tight chemical embrace so the body can get rid of it reminded me of something else I'd read. It had nothing to do with lytico-bodig or cyanobacteria. It was about Lyme Disease, which is caused by a spirochete (or spiral shaped) bacteria transmitted via ticks. It takes a long time to recover from Lyme Disease.

One theory explaining the persistence of Lyme Disease is that the victim's liver processes the toxin made by the Lyme Disease bacteria, then discharges it into the intestines. But, instead of getting eliminated from the body, the toxin gets reabsorbed through the intestinal walls. Patients can't get rid of the toxin because it just keeps recirculating through the body. People who subscribe to this theory have been using an old cholesterol medicine to fight the disease. They say the medicine binds to the toxin in the intestines so it can't be reabsorbed.

Here my brain made a leap that is probably crazy. Lyme Disease comes from a bacteria that produces a toxin. A cholesterol medicine might bind to that toxin so the patient can eliminate it. Fruit bats get a lot of choline in their diet, which humans can get from lecithin. Lecithin also binds to cholesterol. Maybe in some similar way the choline in breadfruit binds to the toxin and helps them control it. Or maybe it just gives their brains more raw materials to manufacture the chemicals a brain needs for proper functioning.

Choline may have nothing to do with the survival of the fruit bats. Vitamin C might not play an important role. Breadfruit might have nothing to do with it. They are merely more clues.

Another interesting thought arose from the comparison with Lyme Disease. The Lyme Disease toxin is manufactured by a bacteria that manages to worm its way deep into body tissues where antibiotics and our natural defenses can't easily reach them. The toxin causes the symptoms, but a bacteria makes the toxin. That is, some bacteria can hide within us, nearly impossible to detect let alone kill. And some bacteria can create toxins that can, in high enough concentrations, cause the exact symptoms of these elusive diseases. Dozens of mysterious diseases, like ulcers, are caused by bacteria. In most cases, this was proven only after decades of lectures by very smart folks who explained why that was impossible.

The lasting result of all this research for me was that I sure wanted to try some breadfruit. I couldn't find any in Colorado groceries, but I did find its close relative, "jackfruit."

Obviously, I bought one.

Jackfruit

I have become a big fan of jackfruit. The flavor reminds me of peaches, but it also reminds me of cantaloupe. I even planted some seeds in a pot in my back yard just to see what seedlings might look like. After a month of watering the dirt and watching, nothing happened so I gave up on that little dream.

What Internet research doesn't tell you about jackfruit is that your hands will get extremely sticky trying to extract the fruit from within its covering. Not sticky like "eat a peach" sticky. This is like dipping your hands into honey.

The goo that protects jackfruit (and breadfruit) is natural latex. Any bat that tried to eat one could not help but swallow some latex. Without doing any more research I decided, based on my own experience, that this latex stuff would stick to any toxins in anybody's intestines, as well as to any bones, sticks or small animals down there.

But I knew even as I tried to wash my hands, that wasn't very scientific. Later I learned that the latex on breadfruit and jackfruit is used to catch birds. Natives coat a stick with the stuff; all they have to do is get close enough to touch the bird and it's snagged.

Micronesian fruit bats also love the nectar in certain flowers. Just like bees, they are indispensable pollinators of some plants. Two of their favorites are the nectar in flowers of the boxfruit tree and the coral tree. I wondered if either of those nectars might contain a mysterious chemical that protected the bats. Reading about the two trees left me more confused than ever.

The boxfruit tree is well named; it's fruit is a little heart shaped box about the size of a fist with a hard shell. The fruit can float for miles across the ocean. When it reaches land, fresh water rain makes it germinate. After a disaster (like a volcanic eruption) wipes out all the life on an island, boxfruit are often are the first trees to begin the natural reforestation. I was looking for some medicinal quality, but it turns out that all parts of the boxfruit tree are poisonous. That's not what I had hoped to learn, but I continued to read. At least the poison in the boxfruit tree isn't useless. Natives all over the world use it to kill or intoxicate fish.

The poisons found in boxfruit are called "saponins." They act like soap; in fact, people have made soap from saponins for hundreds of years. More recently they have attracted attention for use as adjuvants: a chemical that enhances the effect of another one. A much smaller dose of a vaccine provides immunity from disease when administered with the proper adjuvant. Saponins can act as adjuvants. I would guess that, if an adjuvant can amplify the effect of a medicine, it probably could also amplify the effect of a poison. Or maybe of some nutrient that protects a fruit bat. But that's just speculation. I had been looking for a medicine and instead I found a poison, and one that sounded suspiciously similar to rotenone, the fish poison in derris root.

So I turned to the coral tree. We believe the coral tree gets its name from its beautiful red flowers, although its branches can also twist to look like coral. There are over 100 varieties of "erythrina." The trees are quite common in many parts of the world and are prized for their shade. Some varieties live in the southwest United States and in Mexico. In some places, farmers use them to shade coffee plants; in some areas they are used as the framework for vanilla vines to climb. Many of the varieties contain powerful chemicals that are used in folk medicine around the world. One contains an alkaloid also found in the opium poppy. Another contains a chemical that binds to the same places on nerve cells that nicotine does and may influence the activity of acetylcholine. Because acetylcholine seems to be standing near the scene of the crime whenever we investigate Alzheimer's or ALS, that could be interesting. Another has been proven to reduce anxiety in mice, which is reassuring. But all are toxic to some degree and many contain poisons that can kill a human. As recently as 2008, scientists were isolating new alkaloid poisons from these plants. We probably haven't identified all of them yet.

Like so many of the other plants I'd stumbled across that contain interesting chemicals, erythimina are legumes. Their root systems house nitrogen fixing bacteria. Maybe some of the chemicals that make them interesting are actually manufactured by those bacteria. I don't know.

What did this new information mean, if anything?

First, it meant that fruit bats on Guam ate many more toxins than just cycad nuts. Therefore, a nice coconut-fruit bat soup probably exceeded the recommended daily dose of a whole bunch of poisons. Some chemicals are far more effective in combinations than they are alone. This is why doctors warn about drug interactions. A safe dose of drug A taken with a safe dose of drug B can kill you. Maybe some combinations of toxins was dangerous at much lower doses than anyone had considered.

Second, perhaps the medicinal qualities of some of the fruit or nectar was much greater than we know. Maybe those witch doctors and shamans practicing "traditional medicine" knew something we didn't know.

Third, fruit bats eat things that produce adjuvants. Without knowing why the fruit bats don't get sick, this might explain why the Chomorros did. Perhaps soponins that amplify the effect of other chemicals survive to the bat soup stage of the meal. Maybe people in other parts of the world became exposed to both adjuvants and toxins as well. Testing a toxin in the absence of these helper substances might fool us into a false sense of safety about dosages.

Fruit bats make a living by eating plants containing many toxins. Whatever protects them, it doesn't operate within some narrow range. Maybe they simply don't live long enough for BMAA to kill them. But that doesn't explain why these other faster-acting toxins don't.

Why don't the fruit bats get sick? I became frustrated trying to learn about the Micronesian fruit bats of Guam. There just wasn't much information that seemed useful to my particular quest. So I broadened my search to include fruit bats in other locations. Almost immediately, I found a two-year-old study that seemed like exactly what I was looking for.

South American Fruit Bats

Scientists conduct many experiments just to confirm what they believe they already know. It seems reasonable for all oceans to be equally salty, for example, but we can't say we know it without a strict experiment. It turns out, the Atlantic is a lot saltier than the Pacific. Just because the earth seems flat, scientists can't accept that without experimental confirmation. Turns out, the earth isn't flat at all. But we shouldn't say it's a perfect sphere unless we have measured and confirmed it. Turns out, it's not a perfect sphere. Scientists are always looking for assumptions that seem perfectly reasonable to test.

Most of the time, the experiment proves the belief. Scientists get excited when results seem to disprove logic.

In 2008 scientists from universities in Germany, New York, and Boston conducted a study of South American bats to prove a popular common-sense belief. We knew that the bats love to visit mineral springs and "clay licks" These are just what they sound like: bats eat the dirt and lap up the mineral water as if visiting their favorite bar at happy hour and eating the free nachos. We assume they do this to add some nutrient missing from their diet. Why else would evolution favor animals that eat mud? We know the mud has minerals, we know that some diets lack them. But a scientist couldn't say he knew for a fact what was going on without conducting an experiment.

To me, this sounds like some professors from Germany and America deciding they wanted someone else to fund their vacation in the Amazon. But that may just be my sarcastic inner voice. I pictured myself on a grant committee: "You want us to pay how much? So you can fly to South America and watch bats eat mud? Right. We have a special file for proposals like that..."

But these guys were apparently more persuasive than me. And they were really, really scientific. They carefully measured all the minerals in the clay licks. They took tissue samples of bats to find deficiencies. They even used two different kinds of bats, with different diets. One kind of bat eats insects, the other kind eats fruit. What they knew, (but which surprised me) was that the insect diet contained fewer minerals than the fruit diet. The insect-eating bats' bodies had more mineral deficiencies. The insect eating bats needed the minerals in the clay licks. Therefore, obviously, we were going to discover a lot more insect eating bats than fruit eating bats supplementing their diet with mud.

But no. By a huge margin, the customers at the mud bar were fruit eaters. And a high percentage of those were pregnant or lactating females. Their bodies didn't need the minerals. They needed something else.

We know that clay binds to some toxins. Farmers use that information every day. Chemicals they spray on leaves remain potent. What hits the ground is soon inactivated and therefore wasted. And the scientists knew that the fruit bats consumed far more toxins in their daily diet than the other bats. Could they be eating clay to cleanse their bodies of toxins?

A couple of other possibilities sprang to mind while I was reading about this. Perhaps a mineral in the water is useful to bats in some way we don't understand. Maybe one of the minerals the scientists didn't even measure.

Some mineral springs contain lithium in the form of lithium carbonate for example. Beyond elevating mood, for centuries people believed it cured many diseases. It was the basic ingredient in most snake-oil cures of old, those magical medicines sold by traveling hucksters. Many of the popular soft drinks of today were originally forms of "lithia water," including Coca-Cola and Dr. Pepper. 7Up got its name from the chemical formula for lithium. At one point, literally thousands of kinds of soft drinks contained lithia water; some also contained cocaine. Lithium carbonate had good results in a few studies with Alzheimer's and ALS patients. Alas, in high doses, lithium is dangerous, so the government banned it from soft drinks in 1948. Very little research has been done on its properties. Drug companies freely admit they don't study it because it can't be patented. A few suppliers around the country still bottle and sell it. I think I need to try some.

Or it could be another mineral. Some mineral springs have high concentrations of iodine, which has its own interesting history in treating the diseases we're studying. Some have iron, calcium, magnesium, and potassium.

Or, it could be a vitamin. Some studies have shown a correlation between neurological diseases and a lack of B vitamins. But animals can't produce vitamin B-3 or vitamin B-12 and they can't get it from eating plants; only bacteria produce them. Some animals, like cows, have bacteria in their intestines to make B vitamins for them. Once produced, it's used throughout their bodies, but they store it primarily in their liver or kidneys. Meat-eating animals get their vitamins second hand, from the flesh of an animal that stored it. The only other option is to drink stagnant water containing bacteria that produce these vitamins.

Clay licks and mineral springs can provide clay or minerals or vitamins. Unless bats just like the taste of dirt, some biological need is probably getting fulfilled by one or more of these ingredients. We probably need to learn more about each of these, but for right now, let's go back to what the scientists discovered.

Dr. Christian Voigt of the Berlin Leibniz Institute for Zoo and Wildlife Research was in charge of this study. His partners were from Boston University and Cornell. These are not wild-eyed crazy folks but careful scientists. Every scientist writes cautiously, not wanting to make claims he can't support. They know that other scientists will analyze and criticize anything that's less than scrupulously precise. Maybe because these scientists realized they'd stumbled onto something really interesting and possibly important, they were even more cautious in their language, which makes it harder to read. Here's what they said, NOT in their words, but in my own translation into less scientific terms. In effect, they said this:

"Okay, there might be lots of reasons bats eat clay. Different bats might have different motives. Some might need the nutrients. We're not arguing with any other theories here, so don't write angry articles about us. But gosh, guys, in our study, it looked pretty clear: fruit bats drink mineral water and eat clay because it protects them from all the toxins in their diet."

All over the world (but I can't say for sure in Guam) fruit eating bats and birds visit mineral springs and clay licks. Some major tourist attractions are built around them and people come from continents away to watch them. Macaws and parrots are the stars of some South American clay licks. Interestingly, they only visit the clay lick during the part of the year when their only food choice is fruit that contains toxins.

Humans also have a long history of eating clay for their health. It's called "geophagy." Years ago, people from the northern states made fun of southerners for this and called them "dirt eaters." Before you feel too smug, you should know that kaolin clay is the main ingredient in Kaopectate.

Rich people go to fancy spas where they can enjoy mud baths. Elephants and rhinos wallow in mud.

I couldn't help but think of pigs digging up fields with their snouts so they could eat the roots, rhizomes, and truffles they found. I bet they wind up eating a lot of dirt every day. I wonder if the dirt they eat helps them eliminate the toxins that mysteriously don't seem to bother them? Could dirt be the secret ingredient in a pig's diet that keeps it healthy?

Eating Dirt

Madame Line Brunet de Courssou treated ulcer patients at clinics in Africa by feeding them clay, especially a kind of French clay she had to import. Traditional healers all over the world have used clay as a treatment for generations, but Line did something most of them did not do: she kept careful notes about what worked and what didn't. It was almost like a ten year scientific experiment, but without the funding or credentials. She noticed that clay from one site helped, but clay from a different site did not. No one paid much attention to her. She approached the World Health Organization with her findings — fifty cases of people cured by the clay. People were impressed, but she lacked credentials. Before anyone took her very seriously she died.

I suspect there's another movie in here somewhere. The particular ulcer she was treating is caused by a bacteria similar to the one that causes leprosy. The bacteria produces a toxin that does the damage. The clay seemed to kill the bacteria and neutralize the toxin. But that doesn't make much sense, does it? Dirt killing germs? Line's son approached scientists in Europe and the U.S. with her notes. Everyone admitted they were impressive, but they were not scientific. His inquires became more passionate. Finally, an unlikely team of scientists at Arizona State University became interested.

Lynda Williams is a geochemist who studies the chemistry of clay; she would not normally study diseases. Shelley Haydel is a microbiologist who studies tuberculosis, so would not normally have much reason to study dirt. Neither one had the right educational background to tackle a puzzle like this alone. Together they have discovered evidence that some types of clay actually kill germs and destroy toxins. Obviously, this would be cool if further experiments prove it to be true.

The structure of clay is fascinating. One article described each tiny particle as a "sandwich" of a thin layer of one material between two layers of a different material. Because the whole sandwich is very thin, there's a huge amount of surface area. This caught my attention because it was so similar to an analogy I used myself twenty years earlier in my book *There Are No Electrons: Electronics for Earthlings*. Only I wasn't describing clay particles, I was describing electric capacitors.

A capacitor is a sandwich in which two slices of a material that conducts electricity are separated by a very thin layer of a material that does not conduct electricity. It's job is to store static electricity. Capacitors can hold a charge for a very long time. Different materials have more "capacitance" than others and their physical shape affects their characteristics.

I wonder if particles of clay act as microscopic capacitors binding to toxins the same way a speck of paper clings to a comb?

Penicillin

Not many popular movies feature the microscopic war conducted every day between fungi and bacteria. It's an ancient struggle with trillions of soldiers, secret alliances, and battles that change the course of history. For such a movie to be popular, the bacterial battle would have to intersect with some epic human struggle. Luckily, human history was changed by just such an intersection. I think it would make a great movie. If any Hollywood producers happen to read this, here's my pitch:

In the summer of 1933, an absent minded, disorganized 52 year old Scottish scientist named Alexander Fleming was growing bacteria in his lab at St. Mary's Hospital in London. I would suggest some handsome and charming but badly rumpled actor to play Fleming; perhaps Hugh Grant or Johnny Depp. Fleming's got petri dishes scattered all around his lab (and, for the movie, we could throw in stacks of leather bound books, old microscopes, perhaps a few cages with rats and monkeys, and maybe a stern housekeeper who disapproves of the mess.

I'm just making all that up, but you have to admit it makes the scene feel more realistic somehow, doesn't it? It would be reasonable to put some hand blown glass beakers and jars around the lab because we know Fleming liked to do his own glass blowing. We also know he had a wife Sarah, who had been an Irish nurse. I'm picturing Rachel McAdams. Believe it or not, he also had an assistant named Merlin. The lab should appear cluttered: Fleming was not a neat guy; he did not like to throw things away. He enjoyed painting and took that hobby to a new level in a way that says pretty much all you need to know about him. He created little "paintings" by inoculating petri dishes with various kinds of bacteria, each of which would grow into some color of fuzz. By timing his inoculations and placing them carefully, he made little pictures of ballerinas, soldiers, and stick figures. Bacteria was his "paint."

Fleming became interested in things that killed bacteria during World War I when many soldiers died from infections. As our movie opens, Fleming has achieved some notoriety for discovering lysozyme, a natural germ-killing substance found in tears and egg whites. He discovered this by letting some discharge from his own nose dribble onto a petri dish, perhaps to see what interesting colored infestation it might yield for his art. Turns out, something in the stuff killed germs.

He wrote papers about lysozyme in the 1920s and continued to study other things that might kill bacteria. But the Great War had ended fifteen years earlier, those memories were fading. World War II was six years in the future. The world was in the middle of the Great Depression. No one had money for anything, and pure research probably seemed like an abstract luxury. I can imagine a scientist getting discouraged and distracted. I can picture him staring out the window and saying, "what's the point?"

August of 1933 was hot in England and Fleming's lab became increasingly uncomfortable. He decided to take the month off. He just locked the doors and left.

He returned to a whole bunch of petri dishes overgrown with bacteria. Inadvertently, Fleming had created a bacterial paradise; moist, nutrient-rich petri dishes left undisturbed for a month in the heat of a London summer. His inattention was exactly what they needed.

But one petri dish failed to grow bacteria. Instead, it just got moldy. That would have been the first item in the trash for many scientist. Certainly a stern housekeeper would try to wrestle it away from the unkept scientist who never threw anything away. I'm picturing Swedish curses and someone brandishing a broom. Perhaps Rachel McAdams interceded. I'll leave the details to a professional screenwriter.

Fleming's genius did not lie in precision. His genius was that he noticed that one petri dish and thought it was interesting. Some sort of mold had taken hold and the bacteria refused to grow near it. Rather than discard it, he cultivated the mold and experimented with it. He tried (and failed) to isolate the active substance. Finally, he wrote a paper about it. He got it published but no one was very interested in a mold that disliked bacteria. Fleming moved on to other experiments.

Cut to a different laboratory ten years later. This one is nearly the opposite of Fleming's. Everything is spotless, dozens of scientists scurry around in white coats. In contrast to Fleming's casual attitude, everyone in this lab is intensely focused. World War II is raging, England endures nightly bombings. In America, scientists work feverishly to create a nuclear bomb before Germany or Japan do. This English lab has a different mission: they're trying to prevent the death of soldiers from infection.

By this time, we knew that bacteria caused infection. There just wasn't much we could do about it. We could sterilize surfaces with alcohol or heat, but once bacteria (which are also known as germs) started growing inside someone, we could slow them down with sulfa drugs and that's about it. Sometimes it worked. But much too often a battlefield wound became a death sentence. Germs were killing as many Allied soldiers as our enemies were.

Whichever side figured out how to defeat bacteria would very likely win the war. It might prove as decisive as the nuclear bomb both sides were furiously working on.

The scientists tried everything, but nothing worked. They became increasingly desperate, willing to try anything. One started reading the old papers on lysozyme, including the papers Fleming wrote about it. That reminded him of the paper he'd read ten years earlier written by the same guy, the one about the mold. He read it again. Mold did not seem a likely solution, but they couldn't overlook any possibility. It was worth a try, but there are thousands of kinds of mold. How would he ever find the right one in time? He contacted Fleming.

Fleming was happy to help. No, he had not discarded the mold. In fact, he had kept a culture of it alive for the entire ten years.

In the modern laboratory, they grew Fleming's mold and extracted different chemicals from it, trying to isolate the one the bacteria avoided. Finally, one of the chemicals seemed promising. They injected it into half a group of rats, then exposed the entire group to a deadly bacteria. The ones with the mold extract all survived. All the others died. The scientists got excited. They grew more mold and extracted the medicine from it. The process was slow, they didn't have much time and the war raged on. They only had a few tiny vials. They did not know if it would be effective in humans, they didn't know if it might be fatal to humans. But they needed to know.

They decided to try it on a man who was so badly burned on most of his body he had only days to live. His infection was severe, he was in terrible pain. If this new medicine killed him, it might actually be merciful. Either way, they could see if the mold extract had any effect at all.

Amazingly, the extract worked. The infection subsided noticeably. Over the next few days he improved dramatically. Within five days, there could be no doubt. This mold extract was the magic bullet they'd been seeking. The man remained in critical condition, but obviously improving. Unfortunately, they had used up all their medicine.

Without more medicine, the man's infection grew again. He became sicker. In desperation, the scientists processed his urine to recover whatever medicine had passed though his body. They gave that to him and he began to improve again.

But they couldn't manufacture or recover enough. The man died.

The scientists grew mold as fast as they could. They refined their techniques. In a plot twist worthy of any movie, the first person they actually cured of a deadly infection was a five-year-old girl.

But they still could not manufacture enough of this wonder drug. They grew the mold on various surfaces, but the process was too slow. Mold farming was a brand-new activity and they had to make it up as they went along.

Finally, they asked some American scientists for help. Looking at the problem with fresh eyes, the Americans realized that trying to grow mold on any surface limited the yield. What was really needed was something the Americans knew very well: the technology of making beer. Like yeast, fungi can grow in a vat of nutrients without the limitation of a single solid surface. The Americans set up huge vats and started brewing penicillin.

It was still a race. The stuff had to be grown, the penicillin extracted, then sealed into antiseptic vials. But they did it. The first huge batch of mass produced penicillin left the factory on June 6, 1944.

If you recall your history, that date became famous as "D Day," the day Allied troops invaded Normandy to take the fight directly to the Germans.

A Surprise from the Backyard

In some ways, I'm a little bit like Alexander Fleming, but maybe not in any of the good ways.

Yesterday my wife brought a pot of dirt into the kitchen from the backyard. She plunked it down on the counter next to the sink without a word and watched my face for a reaction. It was a cheap, nondescript black plastic pot that had once contained some plant from a greenhouse; a pot we had obviously recycled to grow some forgotten tomato seedling or geranium cutting. My backyard contains several similar pots because, I am told, when I'm done with a project (like transplanting a tomato) I never clean up the ensuing mess. Sooner or later these pots sprout weeds. Sooner or later I clean them up and discard them, sometimes without even being reminded. My first guess was that this time I was going to get reminded. A certain genius-like talent for disorganization is what I share with Fleming. This talent may lose some of its luster for even the most understanding of wives after a few decades.

One sturdy little weed grew in this pot. It was about two inches tall with a thick stem and glossy leaves just beginning to unfurl. I know every variety of weed that's likely to spring up in my backyard. I've used most of them in salads or tea. But I'd never seen a weed like this. I stared at it for a minute; my wife patiently waited for me to understand.

"You're kidding," I said at last and she nodded.

A jackfruit seed had sprouted. We were probably the only people in Colorado, maybe the only people outside Florida in the U.S. to own a jackfruit tree.

Owning a tiny jackfruit tree in a pot is unlikely to lead to any great scientific breakthrough. Still, it felt pretty cool. Appreciating weird little things like that improves a guy's life. But most progress occurs when people read the hard books, conduct the boring experiments and keep careful records. It's not my strength, nor what I enjoy, but sometimes you simply must grit your teeth and do your homework. Before I got distracted by learning about penicillin and jackfruit trees, I was trying to learn more about the toxin that cyanobacteria produce. BMAA can, apparently, cause neurological diseases if you get enough of it.

I decided to see what sort of poison BMAA was. How does it do its damage?

It turns out, it's dangerous in at least two different ways. It can act as a "genotoxin," which means it can alter DNA or RNA. Genotoxins can cause cancer, so that's usually why we study them. Because ALS or Alzheimer's aren't forms of cancer, that aspect of BMAA has been largely ignored. In August of 2011, Peter Spencer proposed that the genotoxic property of BMAA might be much more interesting than anyone ever imagined. We'll get to that later.

Most people focus on BMAA's action as an "excitotoxin."

Excitotoxins

Nerves communicate with each other by using chemicals. One nerve produces a tiny amount of a chemical, the next nerve senses that and responds. It's sort of like using basketballs and hoops as electrical switches. One cell tosses a chemical basketball; if it swishes through the next nerve's hoop, that nerve turns on for an instant. Once the basketball falls through the hoop, the nerve turns off.

Excitotoxins are chemicals that happen to exactly fit in those baskets. The problem is, when they land in a hoop, the nerve turns on but doesn't turn back off. It's like the basketball got stuck in the net. Unfortunately, nerves are designed to flicker on and off. When they stay on, they burn out. Excitotoxins kill nerve cells by burning them out.

BMAA is an excitotoxin; that's what originally made it interesting to scientists. If it can kill nerve cells (which it can) they thought, perhaps that's what's going on in ALS and Alzheimer's. Maybe if you kill cells early in life, your brain will compensate for a while, but at some point you won't have enough functioning brain cells. Maybe, they thought, that's why these diseases make people seem prematurely fragile or senile.

There are dozens of other excitotoxins. Our bodies normally produce glutamate, which fits exactly through those basketball hoops. Our brains use it all the time. Still, if it reaches very high concentrations, it might burn out nerve cells by, in effect, tossing balls through the hoops too frequently. Many scientists are working on things to reduce the excess glutamate in the brains of people with ALS or Alzheimer's.

Other excitotoxins are found in various kinds of food. We ingest them naturally all the time.

Some excitotoxins are used by the food industries. Not only do they excite our brain cells, they also excite our taste buds. Monosodium Glutamate, or MSG, (originally identified in kelp in 1908) is an excitotoxin now added to many foods to enhance flavor. Others include "hydrolyzed vegetable protein" and aspartame (the artificial sweetener sold as NutraSweet and Equal).

If excitotoxins are to blame, why doesn't everyone get sick? That's an active question today. People worry about them, others believe that worry is unfounded. They occur in our brains normally, we eat them routinely. Tests demonstrate that they're safe. Still, some argue, they may reach unsafe concentrations within us. Maybe adding them artificially worsens their effects. Maybe they become more dangerous if our bodies are low on antioxidants. Maybe some people become sensitive to them, especially as they age. Maybe they're more dangerous in liquid form. Maybe they open the blood brain barrier and therefore become more concentrated in our brains than nature intended. The food industry has produced thousands of tons of excitotoxins and we've gobbled them all down. Because they're in so many of the processed foods we eat, they're hard to avoid.

Some people believe that excitotoxins are the biggest culprits behind ALS and Alzheimer's. We consume excitotoxins every day, from the aspartame in diet soft drinks to the MSG in everything from canned soup to salad dressings. I couldn't make up my mind and decided to just be cautious. Our bodies can probably deal with reasonable quantities with no problem. But it's probably smart to limit our intake, just as we do with sugar or salt or any other chemical. Anyway, chemistry in general makes me nervous.

On the other hand, I couldn't stop thinking about the mold that killed Fleming's bacteria. I've always maintained pets, plants, and fish. Living things seem more up my alley. Maybe I needed to learn more about mold and mushrooms and all the other fungi. Maybe there was a clue in there I could follow.

OK, who was I kidding? Sure, there might be a clue, but I had no good reason to think that. What was really going on is that I'd been forcing myself to sit through a sort of chemistry class, learning words like excitotoxin. I knew I needed to learn more about toxins, because they seemed important to the puzzle. But I'm more interested in bugs, critters, and plants. To a guy like me, fungi seemed pretty exotic, with their wild shapes and weird reproductive habits. This seemed like a great time to learn a little more about them. It was more like recess than work.

As it turned out, thinking about fungi led me to thinking about fungicides, which led to one of the most interesting connections I stumbled across during this project.

Fungi

For the first two billion years or so, life on earth consisted of bacteria and things very much like them: tiny, simple life forms. They had several things in common.

A thin membrane encased their bodies that allowed nutrients to enter and waste products to exit. This "differentially permeable" membrane became a standard feature of most life forms, including humans. Many also had a sturdy cell wall made of either chitin or cellulose. What they did not have was the kind of internal organization we associate with modern plants and animals. Their DNA was not nicely packaged within a nucleus, for example. Those that could draw energy from sunlight did not contain neat little bundles of chlorophyll to do the work, like the chloroplast of modern plants. They did not have neat little internal bodies like mitochondria to create energy from food.

All these very simple life forms (like bacteria) are called prokaryotes. All the more complex life forms (like plants and animals) are called eukaryotes.

About a billion years ago, the prokaryotes began to diversify. Some of the prokaryotes that used cellulose as the material for their cell walls changed, over time, into plants. Others remained pretty much the same and survive today as bacteria. Some of the prokaryotes who built their cell walls out of chitin changed, over time, into animals. (Animal cells do not have cell walls, although some still use chitin. The exoskeleton of insects and crustaceans contains chitin). Others evolved into fungi. Fungi still have cell walls made of chitin. With so many varieties of microscopic life, the composition of the cell wall is one of the simplest ways to divide them into groups. If the cell under your microscope has cell walls made of chitin, you might guess you're looking at some variety of fungus. If it has a cell wall of cellulose, it's probably a plant. If it has no cell wall, it could be an animal cell. That's a starting place. Once you get beyond that, it's easy to get lost amid the astonishing diversity and sheer numbers.

Fungi are one of nature's success stories. The class includes mold, fungus, and mushrooms. You might live your lifetime without ever taking a breath that contains no spore from a fungus. You might fear a cobra's venom or anthrax bacteria, but the toxin produced by some mushrooms is deadlier. They can digest wood or hay or the clothes in your closet. They can live in your freezer or your hot water heater or furnace.

Fungi and bacteria evolved in the same primordial soup, sometimes in fierce competition. Some fungi produced powerful chemicals that killed bacteria. Today we call those toxins antibiotics. Penicillin was the first that humans discovered. The mold in Fleming's laboratory typically grows on bread. Today we make penicillin from a similar mold that grows on cantaloupe.

We already know that bacteria aren't shy about fighting back. Some produce plenty of their own chemical weapons. If we haven't even named ninety percent of the varieties of bacteria, it's safe to assume we don't know all the chemicals they produce. Beyond poisoning their enemies, they can reproduce so quickly that sometimes they just devour all the food a fungus, or any other competitor, might want before it can get going.

Other fungi and bacteria adopted a different strategy. They decided to cooperate. Lichens are one example. Some lichens are a fungus and a cyanobacteria living intertwined, symbiotic lives. The fungus builds a structure with its tough chitin cell walls and collects water and minerals (often from a rock), while the cyanobacteria uses the sunlight to provide the energy for both. Most plants use a network of fungus around their roots called "mycorrhiza" to mine minerals from the soil for them. Without these fungus, trees and other plants don't grow very well.

Man has known about fungus since prehistoric times. We've employed them as medicine, as hallucinogenic drugs, and to poison our enemies. Claudius, one of the Caesars of the Roman Empire was nudged off this mortal coil by mushroom toxins his wife Agrippa got from the witch Locusta. Some fungi can "eat" oil from a leaky oil well. Some can digest newspapers and cardboard. Some can gobble down toxins deadly to humans without a hiccup. We've even grown mushroom as food. By 1894 one mushroom farm in England produced 20,000 pounds of mushrooms every year.

But back then we didn't understand much about them. We suspected that the tiny spores they produced were some sort of seed but we didn't understand how they worked. A book of that era flatly stated that no man had ever grown a mushroom starting from a spore. That's because they didn't know every mushroom started from two spores, much as many animals come from a sperm and an egg. Today, raising mushrooms is both a big industry and a hobby for millions. I recommend Paul Stamets' book, *Growing Gourmet and Medicinal Mushrooms* and Rush Wayne's book *Growing Mushrooms the Easy Way* if you'd like to try it for yourself. We now know how to start from spores, but the majority of growers start from tissue cultures.

The last decade of the nineteenth century seems very distant to us today, but consider how advanced other sciences were by 1894. This was half a century after we'd figured out that bacteria caused infections. The Industrial Revolution had transformed the world. No one could remember a time before trains existed. Only a dozen years later, Einstein would publish the papers that would revolutionize physics. The previous year, Tesla had lit the Chicago World's Fair with electric lights. People had been fooling with ideas for making radio practical for decades. How could it be that, in 1894, no scientist had ever grown a mushroom from a spore?

Because we didn't know the science of their lifecycle.

Today we know more, but much remains mysterious. Picture a mushroom that you might buy at a grocery store: a little stem topped by a round cap. The underside of that cap contains tiny balls, each smaller than a spec of dust. One mushroom might produce a million spores. They don't just drift out; each one is shot out as if from a gun. In fact, we can put a sheet of paper beneath a mushroom cap and a day or so later the paper will be covered with spores as if spray painted. The "spore print" we just made can help to identify the variety of mushroom. Obviously, if the spores are white, they will show up better against dark paper. And it wouldn't hurt to cover the whole business with a jar so air currents don't interfere. Different mushrooms produce different colored spores and unique patterns in a spore print.

If one of those spores lands in a nice, moist environment its cells begin to divide. It may grow into a tiny white hairlike form. At this point, however, it isn't acting like a seed sprouting or a fertilized egg hatching. It's fragile. Its growth remains timid. That's because, genetically speaking, it's only half a mushroom. It's more like a sperm or an unfertilized egg. If it is very lucky, another spore landed in the mud nearby. If these two manage to find each other in the soil and combine their genetic material, then they begin to grow fantastically. A single delicate microscope hair becomes a white web of fibers spreading underground, sometimes covering an acre or more. This white web is called mycelium. When Einstein was young, mushroom farmers had to collect or purchase blocks of mycelium to start their crops. I read one little book written in the late 1800s devoted to making bricks of manure for this purpose. It sounded fun, but I decided against it as a career choice.

After the mycelium consumes all the readily available nutrients in a growing bed, growers expose them to light, or increased humidity, or reduced carbon dioxide to trigger them to create "fruiting bodies" which is the part you buy in the store.

Mushrooms produce so many spores because they simply can't compete with their more ancient and primitive kin. Bacteria and simpler molds employ a very basic strategy: they take in nutrients, they grow quickly, they reproduce wildly and they aren't above releasing a toxin or two. A few mushroom spores might get lucky and land in a pile of horse manure and survive the competition to make fruiting bodies. It happens; we find wild mushrooms on farms and ranches and deep in the woods.

photo by Zoran Miljković

But, in a moist and dirty barn, even with an optimistic mushroom farmer, they don't stand a chance. A million mushroom spore can't compete with a dozen spore from "forest green mold," for example. The mold will have consumed everything in sight long before the mycelium can grow strong enough to make fruiting bodies. Even if they make it that far, you won't get enough mushrooms to justify the effort.

The solution is for the mushroom farmer to kill off the competition. First, he sterilizes some appropriate mycelium food (usually agar in a petri dish). And I don't mean casual sterilizing like you might with your baby's bottle or food. Mushroom professionals cook the stuff in pressure cookers at temperatures above boiling for twenty minutes or more. Once cooled, they add spores, perhaps from a spore print. They use special HEPA filters to exclude competing bacteria and mold spores from the air in their laboratories. They seal the lid on the petri dish with tape.

If they've done everything well, a white fuzz will appear on the agar within a few days: new mycelium. If they've done everything perfectly no other mold or growth will appear. If they haven't done everything perfectly, they discard it.

Then they pasteurize the medium they want the mycelium to consume: a bag of sawdust, or manure or old newspapers depending on the variety of mushroom. Pasteurizing is not sterilizing, by the way. To pasteurize something you heat it to about 160 degrees for twenty minutes or more. That kills the bad bacteria and mold, but not some friendlier varieties. You can't prevent a pile of horse manure from getting contaminated by one stray spore. So you're likelier to be successful if you pasteurize rather than sterilize at this point. The few harmless good critters that survive will destroy those few stray contaminants. When you compost horse manure mixed with straw, the composting process itself pasteurizes the mixture. When you're done, you've got a rich universe of living things, but the worst competitors for your mushrooms have been eliminated.

Some mushrooms grow on wood or sawdust rather than manure. One can sterilize or pasteurize sawdust, but there's always the chance some unwanted spores will infiltrate and compete with your mushroom crop. A fellow in Oregon, Rush Wayne, adds an interesting twist to the process. He uses a very weak solution of hydrogen peroxide to moisten the medium. Hydrogen peroxide kills spores, but mycelium can live in it. If you only add mushroom mycelia, you're unlikely to get mold. Using very weak hydrogen peroxide extends the "life" of the substrate, sort of like a preservative would.

Dr. Wayne has hordes of fans and followers, but traditionalists remain skeptical. His way is just not the way they've always done things. They prefer using steam. It feels oddly comforting that the power behind the Industrial Revolution, steam, remains king in the world where ancient life forms meet modern microbiology and capitalistic agriculture.

The science of mushrooms has come a long way. I've raised a few mushrooms myself in my basement, starting from mycelium in a petri dish. I've started mycelium from spores and my refrigerator contains a mayonnaise jar full of oyster mushroom mycelium floating in broth, waiting for my attention to swing back their way. A hundred years ago, Einstein couldn't have done that. Yet we've only scratched the surface. There are thousands of varieties of mushrooms; we can grow only about forty of them with a reasonable shot of success. As to the rest, we just don't know enough about them yet to grow them in any quantity. And it's not because it wouldn't be worth our while. Dried morel mushrooms sell for $240 per pound as I write this. Fresh white truffles go for nearly ten times that much. We still use pigs and dogs to find them in the woods because we haven't figured out how to grow them commercially. A number of individual antibiotics (antibiotics come from fungi) have sales of over a billion dollars per year. Other molds and fungi cost farmers billions of dollars every year.

One kind of chytrid fungus, *Batrachochytrium dendrobatidis* (for some reason, everyone calls it Bd), is believed responsible for a huge and mysterious die-off of amphibians. Another kind of fungus causes the White Nose Syndrome in bats. Millions of bats are at risk of dying from this disease. Bats are so important to agriculture, both as insectivores and pollinators, this epidemic could be expensive to humans. Gwynne Domashinski became convinced the two fungal diseases were related somehow and tried to generate interest in that idea. She ran into some resistance, perhaps because she was only ten years old at the time. But DeeAnn Reeder, a bat specialist with Bucknell University listened. Turns out, Gwynne was onto something. Now in eighth grade, she's been working with the University for four years as part of their research team. When you and I start whining because we don't have the education or background to learn about this stuff, we'd be wise to remember Gwynne.

There's money in fungi and most of the science has yet to be discovered. The biggest mystery is why more young people don't go into the field. For my purposes, the idea of a vast army of living things just as varied, complex, deadly, and ancient as the cyanobacteria they compete with intrigued me. Bacteria and fungi have been battling each other for a billion years or more; other times (as in lichen) they work together as partners. This only lends drama to the mystery. We understand a tiny fraction of either group's secrets, most of which were discovered within my lifetime. How could a person remain incurious?

What Happens to Toxins?

Nothing lasts forever, including toxins. I wondered, what eventually eliminates them?

Some of the poisons we spray on crops to kill bugs and weeds become harmless in the soil. The clay molecules cling to each toxin molecule so tightly it becomes practically inert. We don't worry about them getting into our water supplies or traveling up the plant roots into our vegetables. Heat destroys some, a few days of sunshine breaks down others, like rotenone.

Our bodies produce antitoxins, given enough time. If we got tiny doses of rattlesnake poison over a long enough time our bodies might become "immune" to them. Alas, typically our first dose is a big one. Our doctor will inject some antivenom that an animal has made in response to small doses. Ancient kings and other paranoid people sometimes ate tiny amounts of poison with every meal so they would develop a tolerance for it. More than one paranoid king surprised an enemy by surviving a surefire poisoning plot.

In the 1980s, people got the idea that a big jolt of electricity destroyed snake venom, a nerve toxin. Carrying this idea to a weird conclusion, hunters and fishermen started buying stun guns to take along on their adventures. If a snake bit them, no problem. They'd just shock themselves healthy with the stun gun. The companies making stun guns saw a new market and started advertising in fishing and hunting magazines. The magazines, by sheer coincidence, began running articles about stun guns.

It's a cool idea. There just isn't any evidence that it might work. As recently as 2008, the School of Pharmacy at Oklahoma State University conducted a review of the literature just to make sure. They determined there was no evidence supporting the practice.

Notice that we're talking about toxins, which are poisons, not the bacteria or fungus or plant that produces them. Oxygen can kill cyanobacteria, but it has no effect on the toxins that remain behind. Boiling might kill a hemlock plant, but you would be very foolish to drink the water you boiled it in. Boiling will kill most kinds of bacteria but, if they are full of toxins, boiling just releases the poison into our teapot. Antibiotics are dangerous to bacteria but don't destroy their poisons. A bacteria is like a rattlesnake; the toxin is the venom. Unlike rattlesnakes, cyanobacteria produce many kinds of toxins. We know of about eighty varieties of toxins produced by cyanobacteria. The one we've been focussing on is BMAA. We know that the one variety of cyanobacteria in the roots of cycad trees manufactures this toxin, but what about all those other thousands of kinds of cyanobacteria? Do the ones in my back yard pond make this stuff too?

Paul Cox (the ethnobotanist who came up with the fruit bat idea) and some other scientists conducted a study to determine how many varieties of cyanobacteria make BMAA. Some produced lots, some made only a trace. But 95 percent of the varieties tested made at least some. If you're a little ocean critter eating cyanobacteria, the odds are that you're getting at least a dose of something not so good for you. And just because you avoid Guam does not mean you're safe. Cyanobacteria live in your town; some of them might be manufacturing poisons while you sleep. In 2008, M. Esterhuizena and T.G. Downing (two scientists from South Africa) published their findings about fresh water cyanobacteria. They found that 96 percent of the varieties they tested produced at least some BMAA.

I wasn't particularly concerned about some recognized poison like hemlock, but I wondered about my drinking water. If there was a cyanobacterial bloom upstream from my kitchen faucet, and all the bacteria died, what happened to all those eighty varieties of toxins they made? As their microscopic bodies break down in death (a process called lysis) the toxins escape into the water. But then what happens to them? If they are more dangerous when inhaled, was stepping into the shower a bigger risk than drinking unfiltered tap water? If I was getting daily doses, were they building up in my body like they do in fruit bats, or was my body creating "antivenom" against them?

Obviously, lakes recover and the fish return; something must destroy the toxins. I decided it was probably safe to take a shower. My wife thanked me. After a bit more research, I learned two things that destroy cyanobacterial toxins.

In a lake, pond, ocean, or horse trough, nature provides an unsurprising solution: other bacteria eat the toxins and digest them. In the process, they transform the dangerous chemicals into harmless waste products. I made a note to see if I could find out exactly what kind of bacteria these are. Can they survive inside a human body? If I've got 2,000 kinds of bacteria living in me anyway, why not add one that could pull its own weight? Could we at least cultivate them and add a bucketful to any water supply with an algae bloom? Or do they already do that? If so, is it too late to buy shares in the company?

Within humans and animals one specific group of enzymes detoxifies these poisons. I'd never heard of "GST" but it is one of the most abundant enzymes within our bodies. I breathed a sigh of relief. We have a natural defense. This very moment, your body is manufacturing an enzyme that detoxifies poisons.

The bad news was this: some people develop deficiencies of this enzyme, especially when their diet lacks a couple of nutrients. The one nutrient I could pronounce was "zinc." I wondered if there might be a correlation between the diseases I was learning about and a deficiency of GST? Beyond a lack of, say, zinc in their diet, other things can cause a deficiency. In times of stress, our bodies send all our GST out into our bodies to fight the threat. Once it's been deployed, it takes us a while to build up our reserves again. This might be why we seem to get sick after some traumatic event: the stress exhausts our defenses, then a germ or toxin invades us before we have time to rebuild.

I bet hiding from Japanese soldiers in the forests of Guam involved some stress. I bet starving people often develop nutritional deficits.

We know two things for sure (and suspect several others) that stimulate our bodies to produce this enzyme. They are cheap and readily available. I now eat them every day, just to be safe. It turns out that scientists are feverishly investigating these in the treatment of several of the diseases we're studying. Their big disadvantage is that they're too common. No drug company can patent them, so they can't make money on them. If only they can change one or two molecules and give them a new name, then they'll have a valuable product. That's what they're trying to do.

Turmeric plant. Photo public domain.

GST

One of the most abundant enzymes in a normal human is Gluta-thione S-Transferase or GST. It's in many of your organs. One of its jobs is to track down toxins and destroy them. I first heard about GST while searching for "what in nature destroys the toxins produced by cyanobacte-ria?" The answer to that question, according to all the smart computers of the Internet, is "some bacteria" (which actually eat the stuff and render it harmless by digesting it) and "GST." In a pond covered with scum, certain helpful bacteria help break down the toxins. In the human body, we've got so many bacteria we don't know what they all do. It's possible some of them eat toxins. But we've also got GST. Luckily, normally GST does a great job.

But even a healthy individual can get too much toxin in one dose for GST to handle. If you eat some shellfish loaded with saxitoxin (which the shellfish accumulated by eating the cyanobacteria and dinoflagellates that manufacture it) the bad guys overwhelm the good guys.

During acute stress, your body sends all the GST it has out to defend itself from whatever danger caused the stress, depleting your reserves. Your body doesn't make fine distinctions about this. Maybe you got bitten by a poisonous snake, maybe a saber toothed tiger chased you. Or maybe you just lost your job, the creditors are calling, and you might lose your home. It's all stress and your body only knows a few ways to react. One of these is to pump GST out to fight the threat. Once the threat subsides and the tiger has left, your body gets busy making more detoxifying enzymes. But it takes time. Until you've had time to manufacture more GST, your body is vulnerable. Perhaps this is one reason people seem to get sick right after they've suffered extreme stress. You can think of examples from your own life.

Your body requires certain raw materials to manufacture this enzyme. If you don't have any one of these ingredients, you can't create it. One of the most common deficiencies that limit GST production is zinc. People who are zinc deficient can't make GST.

One of the mysteries of lytico-bodig is why some members of a family got it and others didn't. It is conceivable to me that a deficiency of GST might explain some of that. I can imagine everyone in the family eating some fruit bat soup. Even if the toxins in that soup caused the disease, maybe the people had different supplies of GST in their bodies, so some could detoxify themselves and others could not. It's easy to imagine different amounts of the enzyme within members of a family. Some people feel more stress than others and some find themselves in stressful situations that others avoid. Some people like oysters (a good source of zinc) while others in a family might hate them.

Another mystery is why the disease can remain dormant for so many years after exposure to the toxin (or whatever it is) on Guam before suddenly becoming active? It would be interesting to consider whether or not the GST in a human can fight the toxin to a standstill for a long time, but then, given some additional stress on the enzyme, suddenly it's overwhelmed. Like, for example, if someone went on a diet and became zinc deficient so that they could no longer produce enough GST. Or, if some other crisis depleted the protective enzyme and gave the toxin free rein. I have seen no studies about this, it's just a brainstorm on my part. The disease arises from some cause on Guam and then become dangerous years or decades later.

If any or all of these diseases (ALS, Parkinson's, Alzheimer's, etc.) are caused by toxins, and GST eliminates the toxins, it would be handy if we could encourage our bodies to make more of it. Even if it isn't involved at all, it makes sense to make sure your body has plenty. At least two common substances do this, although that's not what either one is normally used for.

One of these is called DHEA. Your body makes DHEA as a precursor to several other hormones, most notably testosterone. As we age, we make less of it. DHEA has been shown to encourage our bodies to make GST but, oddly, not until you've taken it every day for two weeks. Every store that sells vitamins sells DHEA.

The other GST stimulant is curcumin, a natural compound found in abundance in the spice turmeric. Turmeric comes from the massive root (rhizome) of a plant in the ginger family that humans have grown for centuries. It doesn't have much taste, just a mild earthy flavor, sort of like tree bark. Its claim to fame is that it gives a yellow color to other ingredients, like mustard and curry powder. It's even used to dye some clothing yellow, although it fades in sunlight. It's been used as an antibacterial in Indian folk medicine. It seems to have anti-inflammatory properties. Turmeric is widely consumed in many parts of India. Interestingly, the area where cooks use turmeric the most is famous for having the lowest incidence of Alzheimer's in the world.

We know that turmeric stimulates the production of GST. It's been consumed for thousands of years and doesn't seem to have any negative side effects. If a drug company had discovered and patented turmeric we'd all have prescriptions for it by now and would be paying three dollars a pill. Unfortunately, it's cheap, easy to grow, and in widespread use. What good is a drug you can't make money on?

So the drug companies are feverishly trying to invent artificial turmeric. I'm not making that up. They are spending millions of dollars, very quietly, trying to create a synthetic turmeric. When they do, it will be a compound similar enough to curcumin to do the same thing, but different enough to be patentable, that can be manufactured cheaply in huge vats in some friendly country where the wages are low. The company that succeeds will make a killing.

We'll know someone has succeeded when we start seeing vaguely worded articles about the possible health risks of turmeric itself. Drug companies will release inconclusive studies that "suggest" dangers of eating this ancient spice. If we want the benefits of turmeric, we'll all feel a lot safer just buying their new drug.

Different Ways to Play Detective

A good detective might start investigating a homicide by tracing the path of the bullet backward. It entered the victim's body at this angle, so we trace that trajectory backward and discover that it came from the fourth window on the second floor of the building across the street. From that location, our detective might search the window itself and then the room it looks out from. Then he might examine the hallway that leads to that room and investigate who had access to that hallway. Many crimes get solved that way.

An equally good detective might use a completely different strategy. He might start with the fact that the murder weapon was a gun and ask one question: Who owns a gun?

If everyone in town owns a gun, that approach might be a waste of time. But if only ten people do, the investigation might narrow the suspects. If we know the murder weapon was a hunting arrow, and only one person in the state owns a hunting bow, the police will probably decide to question him first.

Similarly, we've been looking at what toxins can cause symptoms of these diseases. Some of the toxins are pretty rare, so they don't seem like reasonable suspects for most of our victims. Many people come down with ALS in New York, for example, yet few folks in New York eat fruit bats. Maybe they have BMAA in their systems, perhaps from cyanobacteria. One approach would be to track every possible way these folks in New York might have been exposed to cyanobacteria. I think that's an interesting investigation. But maybe there are other approaches.

In the example of our homicide victim, we might start by fine-tuning our definition of "murder weapon." Our victim wasn't really killed by a gun, was he? No, he was killed by a bullet. If you know what you're doing, you can fire a bullet from a length of iron pipe. You can fire one from a campfire if you don't care about accuracy.

If we assume that the "weapon" in these diseases is some toxin, we might widen our list of suspects. We need to do two things.

1. Consider different methods of "firing the bullet." That is, how else might we become exposed to toxins?

2. What toxins are more common among our victims. Could any of them be viable suspects as well?

Inhaling Poison

I read several studies suggesting that some toxins were more dangerous when inhaled than when eaten. I also watched a video about Peter Spencer in which natives applied crushed cycad pulp to open wounds as a folk medicine. Both these processes seemed likely ways to get toxins into a person's body.

The idea that toxins inhaled through the nose might be more dangerous than ones you eat got my attention. Like everything else, I was late to the discussion. In 1999, C.H. Hawkes of the Essex Neurosciences Centre published an article in the Oxford Journal of Medicine proposing that Parkinson's disease should no longer be considered a neurological disease but rather an olfactory disease. We've known for a long time that Parkinson's patients have an impaired sense of smell. Dr. Hawkes maintained that the olfactory region of any brain conducts a lot of non-smelling business. In some animals (I'm guessing bloodhounds fall in this group) two thirds of their brain's activities occur in the olfactory region. In humans, it's a much smaller area and we make that region do so many other tasks that we've lost our acute sense of smell. Dr. Hawkes suggests that the olfactory region is where the toxins we're concerned about probably enter the brain. He made a strong case for his opinion. Interesting.

As I read about this, I realized that I wasn't really new to this conversation. I'd written an entire novel in the early 1980s built around it. Although that novel remains unpublished, I remember all the research I did to write it. The plot involved people developing a deadly allergy to bee stings, which fascinated me after I read a government publication describing that process.

Despite all their protective clothing, bee keepers get stung every now and then. It's part of the job, yet they don't develop a fatal allergy to bee venom more frequently than the general public. However, members of their families did seem to develop this allergy at an alarming rate.

In the early 1980s, the Department of Agriculture figured out why. When disturbed, bees routinely stung the bee keepers' nylon suits. The stinger didn't reach the human's skin, but the stinger got stuck in the cloth and the little sack of bee venom got stuck there with it. When the bee-keeper came into his house for dinner, he hung up his bee suit by the front door, along with all the little venom sacks clinging to all the stingers. Maybe once a week his wife crumpled up the suit and carried it downstairs to the washing machine. By that time, the venom had dried. When she bundled up the garment, the venom sacks broke apart, releasing the dried venom. As she carried it down to the basement, the wife inhaled the dried toxin. Because she inhaled the bee venom, she developed a deadly allergy. Her next bee sting could prove fatal.

Once the world of beekeepers understood this, they changed their procedures and eliminated the problem.

For thirty years we've known that certain neurotoxins can be far more dangerous when inhaled than when ingested. And it's not just bee venom.

The herbicide glyphosate, usually sold under Monsanto's trade name "Roundup" is much more effective when combined with a surfactant, so that's how it's usually sold. In that combination, it kills just about every plant it touches (and also some bacteria and fungus and fish). It has the advantage of not being terribly lethal to humans. Less than a fourth of the people who use it for suicide attempts succeed. Laboratory rats can eat a bit every day for two years without getting sick. Because it works so well, it's become the most common herbicide in the U.S. We use about 200 million pounds per year of the stuff.

I found no evidence that glyphosate has anything to do with the diseases I'm trying to learn about. I mention it only because, according to the American Bird Conservancy (which is concerned with the effect of toxins on birds) the stuff is much more dangerous when inhaled. As they report on their website:

"Most toxicity tests cited by industry and the EPA investigate toxicity through oral exposure routes. The toxicity of glyphosate and the common surfactant POEA is much greater through inhalation routes of exposure, which is a likely exposure scenario for humans residing in areas of Colombia. Experimentally induced inhalation of Roundup by rats produced 100 percent mortality in 24 hours."

Paraquat

Paraquat is an herbicide. Farmers use herbicides to kill weeds for the same reason that mushroom farmers pasteurize manure: a field produces more spinach if the farmer eliminates everything else that's competing for sunlight and water.

People don't worry about paraquat too much. Farmers are careful not to get paraquat on their crops, partly because it would kill the plant and reduce their profits. The stuff they spray on weeds is unlikely to accidentally reach produce in your grocery store because they don't harvest the weeds. Like many other toxins, paraquat binds tightly to clay molecules so we don't worry about it as a lingering danger if we spill some or over spray it. Once bound up in the soil, it gradually breaks down into harmless components. If you don't continue to add paraquat to the dirt, it will all be gone in ten or twenty years. It also breaks down with heat, so cooking your veggies destroys it.

The only real problem with paraquat is that it's deadly poison to humans. Two spoonfuls will kill you, but it's a slow death. It may take a month to finish the job. There is no cure or antidote. Your only chance for survival is to eat fullers earth (a kind of clay) or activated charcoal and hope enough of the poison gets stuck to the clay particles before it enters your blood stream. In poor rural countries, taking paraquat is a favorite method for committing suicide. The largest manufacturer adds a bad smell to it, plus a blue dye, plus an emetic to reduce the chances of accidental poisonings.

In the 1970s, the U.S. Government thought it would be a really good idea to spray thousands of acres of marijuana plants in Mexico with paraquat. In 1977 alone, they sprayed nearly 10,000 acres of marijuana plants, plus twice that large an area of poppy plants with a different herbicide. At the time, the "paraquat pot" program must have seemed foolproof. The likeliest outcome was that you'd kill off all the plants, depriving the drug lords of income while saving American hippies from the horrors of giggling on the couch and eating too many snacks.

It seemed completely implausible that the guys with machine guns would simply harvest their plants before the poison killed the leaves and sell the product anyway. I mean, the stuff had poison on it. Who would be mean enough to sell tainted crops just because they could make millions of dollars? And even if they did, those hippies weren't going to eat the stuff, they were going to smoke it. No one had ever suggested that inhaling paraquat was deadly. It just scarred your lungs. Finally, we were pretty sure that heating the poison, perhaps in a smoker's pipe or rolled contrivance, broke it down. Like I said, foolproof.

But the farmers quickly figured out that paraquat only kills the plants while they are photosynthesizing in sunlight. If the sun doesn't hit the leaves, the plants don't turn yellow. So, as soon as the helicopters sprayed the fields, the farmers ran out, quickly harvested their plants, and kept them out of the sunlight. In one year, Mexico exported about 3,000 tons of marijuana to the U.S., perhaps a third of it contaminated with paraquat. Ten years later, the U.S. government was still spraying paraquat on marijuana fields when it found them in the United States.

In all fairness, I haven't read about any dire consequences of the paraquat pot program. I'm sure if there had been, the companies that manufacture it would have let us know. Or else the government would. Right?

The chemical paraquat was first created in 1882 but no one realized it was useful for killing plants until 1952. Today, the largest industrial manufacturing facility in the UK produces about 8,000 tons per year. Not pounds, tons. A smaller plant in China produces about 6,000 tons a year. It's sold in over a hundred countries including the U.S. Obviously, it's a very useful product and the patent has expired so it can be manufactured by anyone fairly cheaply. Because it doesn't penetrate tree bark, it's used to control weeds in orchards, on coffee and banana plantations, and in olive groves. It's used as a defoliant for cotton and hops. It's a lot easier to pick the cotton if you get rid of all the leaves on the plant first. So they simply spray the cotton plants with paraquat before harvest. It's used the same way before harvesting hops to make beer. It's also used to help dry sugar cane, soy beans, sunflower seeds, and pineapples.

You'd think there would be many reports of illness from such a widely used poison, but I really can't say I ran into many. Apparently, people all over the world are scrupulously careful when using it. Even a slow acting poison would have caused problems by now. If a thousand tons of paraquat pot was smoked in the U.S. in one year alone and it was dangerous, you'd think we'd hear of people coming down with mysterious ailments no one could explain. After all, what kind of toxin takes twenty or thirty years to act?

Another herbicide was used during the Vietnam War as part of the U.S. strategy to eliminate hiding places and force rural populations into the cities. That herbicide was called "Agent Orange." Some studies suggest that soldiers who were exposed to this are much more likely to develop Parkinson's Disease. About 20 million gallons of Agent Orange were cheerfully dumped onto the Southeast Asian countryside, so it's a bit hard to track who might have been exposed. Although the science isn't conclusive, the Veterans Administration just accepts the connection: if you served in Vietnam and developed Parkinson's, they'll probably just agree it's a service related condition.

Recently I've started using a weed killer you may have heard of: vinegar. The stuff you buy in the grocery store is a five percent concentration. Amazingly, it kills some very pesky weeds within hours and garden plants even quicker. You just put it in an old spray bottle and carefully spray the weeds. It's really cheap, too, but your garden might smell like pickles for a few hours. Vinegar is also manufactured in ten and twenty percent solutions, which works much better but is hard to find. Your local hardware store probably doesn't sell it. One reason is that the EPA has strict rules regarding anything that's called an herbicide. The active ingredient in vinegar, acetic acid, can be dangerous to skin and eyes at high concentrations but has uneven success killing weeds at lower concentrations. The government worries that people will be careless with anything called vinegar. I bet there's a really interesting story about organic gardeners who'd like to spray weeds with vinegar and the companies who sell more expensive solutions. Some of those seem at least as dangerous to me. But I've strayed from my story.

One little tidbit about paraquat caught my attention. When combined with another toxin, paraquat causes the symptoms of Parkinson's Disease. It seems to be well known in the scientific community, the experiment has been repeated many times. Perhaps because paraquat has proven to be so useful and relatively benign when used properly, this seems to be considered a quirky little oddity, and not the subject of a great deal of concern. The enabling chemical is also considered completely benign and is widely used to control fungus.

Fungi Foilers

We know of over a half a million varieties of fungi; some are hugely beneficial. A few produce antibiotics, others produce mushrooms that delight chefs, some create raw proteins that vegetarians transform into imitation meat. But many fungi are not well loved. They destroy our potato crops, ruin tomatoes on the way to market, and transform strawberries into ugly balls of gray fluff. The varieties of mold that grow behind damp walls can cause terrible allergic reactions and destroy our books. But we can't take it personally. Fungus has been fuzzing up the world for a billion years or more, while humans only invented agriculture 10,000 years ago.

Let's translate that into a scale that's easier to visualize. Let's say a month equals 100 million years. On that scale, fungi stared getting serious about evolving ten months ago. Humans moved out of their caves, gave up the mastodon hunting business, and started planting crops 42 minutes ago. Fungi don't hate us, they've hardly had time to notice we exist yet. We can't be blamed for not understanding them either; on the same scale, we didn't see them through a microscope for the first time until you started reading this page. You were born within the last three seconds.

Before you start thinking that fungus are "ancient," having been around for ten months, consider this: On this scale, cyanobacteria started manufacturing oxygen and brewing toxins beneath the pink sky about three years ago.

Some fungi research focusses on growing them for food or antibiotics. A completely different line of research focusses on killing them so they don't ruin our crops. In 1934, Tisdale and Williams patented a "dithiocarbonate" fungicide. This type of fungicide became widely used within ten years and new varieties continued to be developed into the 1960s. One interesting fungicide is called maneb. It belongs to the specific group of fungicides called "ethylene-bidthiocarbonates" or "EBDC." I think I'll just call it maneb.

Maneb (and its cousins in the EBDC family) sounds like an ideal fungicide. First, it kills the molds and fungi that cause a whole bunch of plant diseases. It protects seeds from getting consumed by hungry molds before you can plant them. It keeps tomatoes, potatoes, and fruit from molding on the way to market. These chemicals increased the usable food supply a lot. Many people would have starved to death without them.

It also seems to have some environmental advantages. It binds to soil particles, so it's not too likely to get into the water supply. It doesn't really dissolve in water anyway. It is only considered "moderately" toxic to lab animals. It breaks down in the presence of moisture, oxygen or sunlight. Heating breaks it down into different chemicals, although some of these are possible carcinogens. Rats can eat much higher doses than the government allows humans. They can eat those doses every day for two years and it doesn't seem to affect them. One interesting and odd bit of information is that guinea pigs can survive three times the concentration that mice can. Birds seem practically immune to the stuff.

But there's some bad news too. It is deadly poison to fish. A concentration of less than two parts per million can kill fish. And, although cooking breaks down maneb, it breaks it down into "ethylenetiourea" (ETU) which we're pretty sure causes cancer and birth defects. So, wash the stuff off your fruits and veggies before you cook them. In 1997, California asked the manufacturer to label it as a potential "skin sensitizer." And, as Kenneth Weissmahr and a team from the University of California at Berkeley reported in 1998 (for the journal of the American Chemical Society), although large quantities of this type of fungicide are released into the environment every year, "little is known about their fate and effects because sensitive analytical techniques are not available."

The most interesting thing about maneb to me was this: if you want your lab rats to develop the symptoms of Parkinson's disease, maneb can help. Adding maneb to either the insecticide rotenone or the herbicide paraquat dramatically increases the chances that your lab rats will get sick. The combination is much more dangerous than either one alone, although I'm not sure we know why.

It occurs to me that a nice unwashed organic tomato could easily have both the insecticide rotenone and the fungicide maneb, because both are permitted under organic guidelines. A tomato from an unknown source, say, some undisclosed Central American country, could conceivably have a trace of paraquat. According to the book *Tomatoland* by Barry Estabrook (2011), a "nonorganic" Florida tomato farmer has over 100 pesticides, herbicides, and fungicides in his perfectly legal arsenal. Although no individual tomato endures all of them and most of the poisons probably don't make it to the final product, his book is a chilling reminder to be careful what you eat. And what you let your workers become exposed to. Even one migrant farm worker giving birth to a child without arms or legs after the mother got splashed with perfectly legal pesticides seems like too many. Don't read his book if you're squeamish. Unless you eat tomatoes.

Detectives Consider Contributing Factors

Let's think about our imaginary homicide investigation again for a minute. Our victim was shot, the weapon was whatever fired the bullet, and the victim did not survive. A smart detective would consider a few more factors: was there only one victim? Or were dozens of other people killed as well? That's obvious in a homicide, but less obvious in a disease. If only one guy was killed, we can look for people with a motive to kill him. If many people were killed, what did they have in common? Were they all just standing on the wrong street corner when some lunatic opened fire? Or can we get a clue by searching them as a group? Did they all belong to some cult? Were they all wearing blue hats? Many Chamorros fell victim to lytico-bodig, while others on Guam weren't affected. That's why they interest us; we hope to locate common factors. Much of the thinking about neurological disease takes this approach: what do the people who contract these illnesses have in common?

Even if our homicide victim was the only person killed, you'd still automatically ask how many gunshots the witnesses heard. A single fatal shot suggests he was singled out by the killer. On the other hand, if the killer sprayed the crowd with a machine gun and our victim was the only guy who fell, we imagine a different story. Maybe the victim was the only person in the crowd who failed to wear his bulletproof vest that day. In that case, his problem was not merely the madman with a machine gun, but that he was vulnerable. He lacked something the rest of the people in the crowd had.

It is possible that some factors protect people from whatever causes ALS and Alzheimer's, just like a bulletproof vest protects policemen from criminal gunfire. If we could find those, that would be excellent progress. Perhaps GST and other enzymes help protect people. Because bacteria digest toxins in nature, perhaps there are bacteria in our intestines that protect us. Nutritionists claim that the intestinal flora of overweight people is quite different from that in thin people. Many people claim that autistic kids support a different mix of critters within their intestines than other kids do. I wonder how many conditions the microscopic life downstream from our stomach affects?

One factor can make another one more dangerous. A firecracker might startle us, but a firecracker plus a stack of dynamite can send us over the rooftop. A virus causes the common cold, but it's worse if you're tired, malnourished, or stressed. Maneb makes paraquat more deadly; maybe other chemicals do as well. Adjuvants make medicines more effective, perhaps some adjuvants make poisons more deadly. Looking for dynamite is smart, but noticing firecrackers would also be good.

Sometimes a deficiency amplifies a problem. It's not safe to drive after a few drinks. But alcohol slows your reflexes more dramatically if you also haven't slept in two days. The guy who drives his car into a ditch might argue that he wasn't "drunk," he'd only had one beer. He may be telling the truth, but it isn't the whole truth. He was also deficient in sleep.

It might be smart to look for contributing factors in these diseases. Could some substances or processes within our bodies protect us? If so, then deficiencies in those substances or processes might be contributing factors.

To illustrate what I mean, let's look at three kinds of substances:

1. Things that help medicines (or toxins) get into your brain, perhaps making them more effective (or dangerous);

2. Things which promote the "good" bacteria in our intestines that might be part of our normal defense against toxins;

3. Vitamins and minerals that play a role in our defense. A deficiency in one of these might increase our vulnerablility.

Mannitol and Inulin

Just like BMAA is an amino acid our bodies don't normally use, our bodies also can't process some substances that are chemically considered sugars. If they taste sweet but our bodies ignore them, we might call them an "artificial sweetener" or "low calorie natural sweetener." Aspartame is such a substance. Although they don't act as fuels for human cells they aren't inert. Some do interesting things within us that you might not expect. One of these is called "inulin" and another is "mannitol."

I stumbled across inulin two years ago in very typical fashion.

I don't like to use weed killers on my lawn or garden, but that means an annual hand-to-hand combat with bindweed and dandelions. I rarely win this battle. I have even been known to get distracted and move on to a new project before the job is finished, believe it or not. Every year the surviving weeds seem to come back better armed, fit, and rested. Dandelions seem especially hardy in Colorado. When my lawn began to resemble a yellow road in the Wizard of Oz, I knew I had to get serious. The neighbors began glancing at me the way the villagers looked at Dr. Frankenstein. I imagined that an awful lot of my suburban neighbors carried pitchforks.

I knew my enthusiasm for digging weeds would falter unless I tricked myself and tricked the weeds, probably with some clever language. I'd never finish the chore of "digging weeds." But I love to garden and harvesting is the best part. I'd heard that some people brew a beverage from dandelion roots. A light flashed in my brain. In an instant I felt myself transformed from "failed homeowner with a lawn that disgraced the neighborhood" to "master urban organic gardener about to harvest a record crop." Weeds are, generally speaking, indestructible. But "valuable natural resources" tend to disappear no matter what you do. I declared the dandelions in my back yard a "valuable natural resource" just to discourage them.

Puffed full of pride, I began digging the roots. I discarded the greens. I know you can eat them. But cleaning and cooking them seemed like too much work.

The roots alone filled three big paper grocery sacks. I cleaned the dirt off them in the kitchen sink. I cut them into pieces about two inches long and ground them up in the food processor. I know what you're thinking. Yes, you're correct: my wife was not home. When the roots were chopped into bits about the size of grains of rice, I dumped some onto a cookie pan, spread them out, and baked them in the oven very slowly at 250 degrees.

Within an hour or so, a most unusual smell filled the house. A lovely, earthy smell, like wet tree bark, or sawdust, or tea, but completely new to my senses. How remarkable, I thought, to have lived more than half a century and then get surprised by a brand-new smell. And to realize that cavemen probably knew this scent.

When the stuff was completely dry, I did another batch, then another. In small dry bits, the whole harvest filled one large coffee can. I put a heaping spoon of it into a teapot and boiled me up some dandelion tea. After some experimenting I added just a pinch of cinnamon, ginger, and cardamom. Just because of who I am, I gave it a name ("Donde Leon"), invented a phony story about its ancient inventor, (a Spanish explorer named Leon) and printed up labels. Ultimately, I gave jars of the stuff away as Christmas presents.

But that may be more than you really wanted to know. Suffice it to say, my family thought it was delicious and I ran out of dandelions in my back yard. Because I was giving this stuff to family, friends, and miscellaneous guests, I started to wonder if there were any health effects, good or bad, from drinking the concoction. That is, having given nearly everyone I love generous samples, it occurred to me to research whether or not it might kill them.

As I suspected all along, it isn't dangerous. In fact, dandelion roots are rich in inulin, a kind of sugar that humans can't digest. It's not particularly sweet to our tongues and it isn't even soluble in water unless you boil it. But, it turns out, it's a "pre-biotic." At first I thought that a typo; don't you mean "probiotic?" But no. Probiotics are all those helpful (in fact essential) bacteria living in your intestines. Probiotics bacteria make vitamins for us, process bad things we eat, and help us digest our food. They are essential to our immune system. In fact, some estimates claim that ninety percent of our immune system is the result of bacterial activity in our intestines. Prebiotics are foods these good bacteria love to consume.

One problem with some medicines that kill bad bacteria is that they can also kill good bacteria. According to Bill Bryson in *At Home* (2010), you're carrying around four pounds of bacteria this moment. Thank you, Bill, for putting that image in our brains. With 2,000 kinds of bacteria battling inside you, scientists don't know what roles they all play; it's hard to fight the bad ones without killing some good ones. But you can at least feed your troops. Inulin is found in many foods. Besides dandelion root, a form of inulin is found in chickory root. It's also in jicama. I now add shredded jicama to my cole slaw, which adds an interesting, slightly sweet flavor and a nice crunchy texture. I don't always explain the health benefits, because many people would rather not discuss intestinal bacteria while dining.

Mannitol is another sugar type of chemical (technically, a "sugar alcohol") that humans can't digest. This one tastes sweet and has the added benefit of absorbing heat when it gets wet. That fine powder on chewing gum and breath mints is often mannitol. It tastes sweet and feels cool on your tongue so it's quite popular.

Mannitol was discovered centuries ago. The flowering ash tree makes it. People originally thought maybe it was the "manna" mentioned in the Bible — a food the Old Testament God gave to his hungry followers as they escaped from Egypt — and that's where the name comes from.

Some of your "good bacteria" might enjoy mannitol so it could be considered a prebiotic as well but that's not what's interesting about it. Although our bodies can't use its calories, it has an even weirder effect: it sucks a tiny bit of water out of some kinds of cells, which causes them to shrink. Doctors and scientists use this property to get other chemicals into your brain.

Normally, your brain is protected from absorbing most of the miscellaneous junk flowing through your blood by the "brain/blood barrier." This is a wall made of endothelial cells (sort of like skin cells) packed tightly together. Mannitol shrinks these cells just enough that a tiny gap forms between them. If a doctor needs to get some medicine into your brain, he may prefer to inject it into your arm rather than drilling a hole in your skull and mannitol helps that plan work. Recently, scientists have been experimenting with injecting blood cells from umbilical cords into patients. These cells are like stem cells; they can develop into other kinds of tissue including, perhaps, cells that have been damaged by Parkinson's disease. They've been using mannitol to get the fresh cells past the blood brain barrier.

Mannitol is sweet, yet indigestible by humans, and it draws water to it. For these reasons, it's often used as a laxative for babies. It's common, it's cheap, it's relatively harmless. That also makes it ideal for drug dealers who want to dilute their illicit merchandise. You make more money on your drugs if some of that powder is really cheap baby laxative. Not only that, by opening the blood brain barrier, the drugs may seem more potent. "Baby laxative" has become common street slang for mannitol used to dilute drugs and maximize profits.

I can't help wondering, though: if mannitol allows umbilical blood cells, medicines and illegal drugs to enter your brain, wouldn't it also allow toxins the same easy access? I'm not suggesting that the mannitol on chewing gum leads to neurological diseases. But obviously some common things do allow unusual access to our tender brains. Just wondering if it might be worth looking at all the common substances that open the brain/blood barrier. Maybe we should notice if we ever eat them along with the toxins we know can cause problems.

Iodine

A few pages ago, we mentioned maneb, a fungicide used to keep many crops safe from mold on their way to market. The idea that a little bit of maneb made other toxins much more dangerous fascinated me, but I had no real interest in learning more about the stuff. Then, by accident, I stumbled across a description of how maneb actually works and got interested all over again,.

In animals, maneb disrupts the ability to use iodine. It may do other things too, but this caught my attention. As a young man, I had a brief love affair with iodine to the point my friends teased me about it. Now that I'm pretty sure I don't have any political aspirations, it's probably safe to share the story with you.

Growing up in Colorado far from any ocean, I managed to live my first twenty-two years without ever tasting squid. It just wasn't available in grocery stores or restaurants out on the prairie at that time. One night, as a young married man with small children, I took my family to a restaurant that had squid on the menu, although it was probably called calamari. (I always tell my sons, "call it squid when you're buying and calamari when you're selling"). I ordered it, liked it fine, and was glad I tried it.

A couple of hours later, back at home, I noticed something weird: I felt really good. Unusually good. Energetic, cheerful, optimistic. I felt so much better than usual that I retraced my day, trying to identify what I'd done differently. The only thing I could think of was the squid. I determined to learn what nutrient was in squid that could affect someone so dramatically. What I discovered was that squid contains a lot of iodine.

Then I discovered that the easiest way to get more iodine into your diet is to eat kelp. Shoot, they even sold little kelp tablets for people who didn't have time to gnaw on raw seaweed or dried squid. I started popping kelp pills whenever I wanted to feel more energetic. The sixties had just ended, other people my age indulged in far more controversial pharmacological experimentation. Some were pot heads, others were speed freaks, snowbirds, and day trippers. I was the kelp kid.

Your thyroid gland requires iodine to do its work. It produces chemicals that help us turn food into energy. If your thyroid is under active, you'll be lethargic and tend to gain weight. My thyroid gland was probably a little low on iodine, so the kelp pills fixed that.

Because the thyroid gland stores most of the extra iodine in our bodies and because it is so sensitive to a deficiency, we tend to think about iodine only in terms of the thyroid. But most of our cells require it for proper function and it's stored in other parts of our bodies as well.

The element itself has a fascinating history.

Between all the wars fought at the end of the 1700s, the world required a lot of gunpowder. To make gunpowder, you need saltpeter which you got from the ash of burned wood. In France, because Napoleon had been so busy, they'd burned up most of the trees and needed a new source. So they tried burning kelp.

It worked fine, but it corroded the pots they used in the process and made them hard to clean. In 1811 a guy added a bit too much sulfuric acid to his kelp ash and noticed a lovely violet vapor rising from his pot. He recreated the situation but this time captured the vapor and collected the crystals it formed. He sent those to the two biggest scientists he could think of, one of whom was Ampere, the guy for whom the unit of electrical current is named. What he had discovered was iodine.

In the early 1800s we discovered how useful iodine was for cleaning wounds. This is because it kills germs, although that concept didn't arrive until fifty years later. People's wounds just healed better if they were cleaned with iodine and we didn't really care why. By the late 1800s it became common to use "tincture of iodine" (iodine dissolved in alcohol) for all sorts of disinfecting purposes. In that form, it can irritate skin and stain your clothes, so other kinds of iodine solutions were developed, including iodol and clioquinol. These were used both as lotions for the skin and medicines to swallow. People took them with the idea of curing diarrhea and other ailments, although it appears that they didn't do much for intestinal problems. Still, if they made your scrape heal faster, it only seemed reasonable they'd make your insides heal faster too so people kept taking the stuff. By the early 1900s, if a doctor couldn't think of anything more specific to treat someone's unusual symptoms, many times he resorted to simply giving the patient iodine. A surprising lot of the time, it worked. They just had no idea why. It became the miracle drug of the era.

For centuries, we've known that a lack of dietary iodine can cause problems. If you are severely deficient, your thyroid gland works overtime to compensate. It swells and you develop "goiter," a massive swelling on your neck. Over a thousand years ago, Chinese doctors treated goiter by giving the patient iodine in the form of seaweed or powdered animal thyroid glands. In the 1700s, people named a condition of extreme mental retardation (with other physical symptoms) "cretinism." We now know cretinism was (and is) caused by a lack of iodine. By 1925, the Morton company started adding iodine to U.S. salt. As a result, cretinism and goiter pretty much disappeared from the United States. But even today, lack of iodine is the primary cause of mental retardation in the world, especially in places where the normal local food doesn't provide enough. Some prenatal vitamins don't contain iodine.

Because salt can contribute to high blood pressure in some people, most of us have tried to reduce the amount we eat. For some reason that I haven't been able to track down, the folks who package salt have also been reducing the amount of iodine they add to it. Most table salt is mined rock salt which has been ground up and sprayed with a tiny amount of potassium iodate. The salt itself might last forever, but the iodine part can evaporate or oxidize away over time. "Sea salt" is the result of evaporating ocean water and contains almost no iodine. "Pickling salts" don't contain iodine because it might discolor your pickles. According to a 2010 study, the vast majority of fast food restaurants use salt that is not iodized. As Jeffrey R. Garber, MD, explained, "...if you relied on fast food, and only a subset of fast food, you could come up being iodine deficient." Between eating more gourmet potato chips flavored with "sea salt" and less iodized table salt at home, and more fast food, we're ingesting less iodine than we used to.

A small group of very vocal iodine proponents feel this is terrible. In the U.S. we get plenty of iodine to prevent goiter and cretinism, but some people argue passionately that we get only a fraction of the optimum amount. In their opinions, iodine deficiency is the root cause of many modern health problems.

Iodine is one of the four "halides." The others are fluoride, bromide, and chloride. They are chemically very similar. Within our bodies, cells have halide receptors; specific places a halide molecule will fit and do some job. The receptors can't really tell the difference between the halides. Any one of the chemicals will fit the receptor. The fluoride and chlorine in your drinking water will take the available spot unless the bromide in your bread does first, or the iodine in your squid sandwich does. Incidentally, until 1960, all bread makers used iodine as part of their process. Then the FDA, apparently nervous that people might be getting too much iodine, made them switch to bromide. Farmers also started using bromide-based insecticides to sterilize the soil before planting certain crops. In Florida alone, they use truckloads of a deadly bromide-based poison to treat nearly all the soil into which they plant tomatoes. Most of the U.S. winter tomato crop comes from Florida.

Fluoride is also a halogen. People started noticing a relationship between fluoride in drinking water and fewer dental cavities back in the 1850s. In the early 1900s they studied the curious fact that some people in Colorado had brown mottled teeth, but no cavities. It turns out that the Colorado water contained lots of fluoride. In the middle of the twentieth century, scientists conducted a study in Grand Rapids, Michigan that demonstrated that adding fluoride to water reduced cavities.

The addition of fluoride to water, toothpaste, and mouthwash has dramatically reduced cavities in the parts of the world that use them. But some people think that all that fluoride also may be finding its way to halide receptors in the body that would really do a better job if a nice iodine molecule stopped there instead. In the 1950s, the right-wing John Birch Society became convinced that fluoridation of water was a communist plot. Of course, they thought pretty much everything was a communist plot. Still, it did not help to promote a calm discussion. Neither did the 1992 outcry from the left wing, which suggested that fluoridation was a capitalist plot to get rid of waste products by the aluminum industry. It's no wonder it's hard to find "facts."

It's a similar story with the other halide, chloride. In the mid 1800s, scientists discovered that water treated with chlorine didn't cause cholera the way untreated water did. This was hard for upstanding people of reason and common sense to accept. Bad tasting water (treated with chlorine) was actually healthier than the good-tasting water that caused disease. But it was undeniable. We know now that chlorine killed the bacteria that caused cholera, but that was pretty advanced science for the time. By the early 1900s, some water supplies were routinely adding it.

Today, it's hard to find water in the U.S. that doesn't contain chlorine. Chlorine and chloride are not identical, but they're pretty close. Chlorine is the poisonous gas we use in dilute quantities to disinfect our water. Chloride is its form when combined with a metal to form a salt (for example, sodium chloride, aka table salt). Either way, we get plenty of that halide as well. While we're brushing our teeth with fluoridated paste and drinking fluoridated and chlorinated water, we're eating a lot less squid and kelp than we used to. Some people believe we're not consuming as much iodine as we should compared to the other halides.

Besides the action of the fungicide maneb in animals, the possible role of iodine in neurological diseases kept showing up in two completely different ways, both of which seemed fascinating.

Since the 1950s, scientists have been discussing the geographic pattern of ALS, Parkinson's and MS. The farther away from the equator you were born, the likelier you are to get all of these diseases. Statistically, that seems to be the case, but that doesn't necessarily prove anything. As early as 1958 people began to set forth theories about each of these diseases that linked them to iodine deficiencies early in their lives. The towns with the greatest incidents of these diseases also had unusually high rates of goiters, which we know is caused by iodine deficiency. There seems to be a definite relationship between low iodine in children and high rates of these diseases later in life. But why would areas farther north (in the northern hemisphere) have less iodine? In 1959, a scientist named H.G. Warren came up with an interesting explanation for the increased cases of MS in northern latitudes: glaciers.

Glaciers scrape the topsoil away from the rock beneath it. When the glacier recedes, topsoil has to form all over again from sand, rocks, and decomposing organic material. Topsoil gets most of its iodine from rainwater, but it's a slow process. It can take centuries to rebuild topsoil to "normal" levels. Until that happens, the soil remains low. The soil near the Arctic Circle remains deficient in iodine, which means the grass growing on it is also deficient, and so are the cows that eat the grass and so is their milk. Babies who drink that milk can't help but be deficient.

The farther south from the Arctic Circle one travels, the more time the soil has had to replenish itself. Whether it's coincidence or not, the farther south you get from the Arctic Circle the fewer incidents of these diseases are reported. Some people say that is no coincidence. They say it's a strong argument that early iodine deficiency predisposes one to these diseases. They believe that babies whose mommies were iodine deficient while pregnant and babies who drink milk that is low in iodine are at a much greater risk of developing ALS, Parkinson's, or MS. That is their explanation for the geographical evidence.

The logic behind this theory seems reasonable and statistics seem to support it, but I can't tell if it has generated any sort of consensus. To "prove" the theory, you'd have to test a bunch of babies, wait for decades, and notice if the ones who were iodine deficient came down with the diseases more frequently. One could argue that maybe it's the number of hours of sunshine in a day during your childhood, not iodine. One could argue that listening to Canadian radio stations is the critical element. Obviously, the iodine fans are not going to agree with those suggestions. But it remains hard to prove.

On a shorter time scale, we know that severe iodine deficiencies in babies leads to cretinism, a form of extreme mental retardation. It doesn't seem like a huge leap to suggest that less extreme iodine deficiency might lead to other neurological problems.

Several of the scientists who support this idea add a little note to their studies that always goes something like this: if iodine deficiency early in life plays a role in these diseases, maybe somebody ought to do an experiment to see if taking iodine would help a patient who has one of them. As far as I can tell, no one has. At least not intentionally.

On the other hand, many people seem to have decided that an iodine deficiency plays a role in diabetes. Thousands of diabetics take extra large doses of iodine and firmly believe it helps them control their blood glucose. The Internet is full of anecdotal stories but I haven't found any real scientific studies that confirm this. Still, because Alzheimer's and diabetes seem linked to each other, it makes sense to notice things that affect either one. Suzanne de la Monte of Brown Medical School calls Alzheimer's "type three diabetes."

The other way iodine popped up also involved Alzheimer's patients. Again, I haven't found many scientific studies that "prove" this, but I found enough stories that it intrigued me. The stories always go something like this:

Grandpa had Alzheimer's. He couldn't recognize his family and couldn't remember simple words. He could hardly speak. Then he got an infection of some sort and the doctors treated that with one of a few specific medicines including minocycline and clioquinol. Within days, mi-

129

raculously, his brain returned. He could recognize people, recall things, maintain a conversation.

The stories often seem to end badly. Once the infection was cured, they took Grandpa off the medicine and the Alzheimer's returned. Or, he died shortly after this reprieve.

Minocycline is an old antibiotic that was once popular for treating acne. It attacks the metabolism of very simple, ancient bacteria, like cyanobacteria. It has proven effective enough times in these odd anecdotal cases that scientists are now doing specific tests with it. Because they don't believe a bacteria is causing Alzheimer's they're looking for some other activity the drug might have. It has an anti-inflammatory characteristic. Some people make the case that it gives the patients' brains a break from inflammation, allowing it to heal a bit. Others suggest that perhaps it escorts metals out of the body, and that ability is responsible for its weird and unexplainable occasional success. Scientists are working on experiments as I write these words to try to answer those questions. In an odd set of anecdotes, some people maintain that minocycline slows the onset of ALS in some people who are genetically inclined to develop one variety of it. But only if administered before the onset of symptoms. Given after the symptoms have started, it seems to actually make the condition worse.

The same story is repeated from time to time, but instead of minocycline, the patient received clioquinol. This is another very old medicine that kills bacteria, but it does it in a different way. Its active ingredient is iodine. It was developed in the late 1800s as an alternative to tincture of iodine and became popular for gastrointestinal ailments. It was used that way, and also in an ointment for skin infections, for over fifty years without causing any particular stir. There's a bizarre twist to the story of clioquinol.

In the 1950s, people in Japan started getting sick from a mysterious ailment. Some became crippled, some went blind. They blamed clioquinol and sued the manufacturer. This was a huge deal; 10,000 people were affected. The company that made the medicine lost the lawsuit and paid out nearly a billion dollars in claims. Obviously, they withdrew it from the market.

But outside Japan, people were distraught that the medicine was going to become unavailable. They said clioquinol hadn't caused them any problems. The government of Egypt took the remarkable step of petitioning the company to keep providing it, saying that if clioquinol caused blindness then half the people of Egypt would be blind already.

What occurred to me is this: there's an obvious reason the medicine was dangerous to Japanese people and not Egyptians. Japanese people eat tons of seaweed in their traditional diet. They get approximately fifty times as much iodine every day as Americans do; Egyptians probably get even less than Americans. The huge amount of iodine in the Japanese diet is often credited as the reason Japanese women have the lowest incidence of breast cancer in the world: breast tissue stores iodine. Adding a whole bunch more iodine in the form of clioquinol to a Japanese person who was already getting fifty times more than an American might well have caused problems. A person can die from drinking too much water or breathing too much pure oxygen; anything you consume in excess is likely to cause problems.

Had the manufacturer bothered to ask me, I could have saved it the nearly billion dollars they paid out to victims. Because I was in elementary school when all this happened, I probably would have been very reasonable.

Just like with minocycline, scientists are currently testing clioquinol but are not especially interested in its bacteria killing properties. Because they don't believe that some bacteria causes these diseases, they're trying to figure out what else the medicine might do inside a human.

If a childhood deficiency predisposes a person to these diseases, doesn't it make sense that a lack later in life might contribute to them? Anyway, I'm curious about *why* a lack of iodine in infancy predisposes people to these diseases, if that's true. Doesn't that seem like a huge clue? If maneb makes paraquat more dangerous to animals, and maneb reduces our ability to use iodine, doesn't it seem likely that an iodine deficiency might contribute to these diseases?

Personally, as an experiment, I stopped using a fluoride toothpaste for a while and instead got a bottle of povidone iodine. I was afraid it would stain my teeth red, but happily it did not. In a strange twist, the health of my gums improved dramatically. On the other hand, not too long after making this change I developed an infected wisdom tooth and needed two root canals as a result of infections beneath my molars. This may not be the best tactic for getting more iodine.

I also decided to find kelp pills, like I'd taken years ago. My local health food store only had pills with a tiny amount of iodine in them. The saleswoman looked aghast when I asked where they kept the larger doses. With a somewhat condescending attitude, she implied that a larger dose would be dangerous, but that I probably would not be able to understand the science behind it.

I removed the other items from my basket, returned them to the shelf, and left the store. As it turns out, one can buy powdered kelp at several other grocery stores in my area. I sprinkle some on food when I feel the urge.

Odd Stares and Iodine

Yesterday morning at the gym, something mildly weird happened. While I manipulated embarrassingly small dumbbells, a guy walked past, very casually, and looked at me. His brief glance sent a little shiver down my back and I wasn't sure why.

I don't recall ever being "afraid" of any human and I wasn't "afraid" of this guy or even nervous. He was much smaller than me, much younger, and not noticeably athletic. He had pale skin with short dark hair. He didn't stare at me or try to make contact. Nothing threatening at all. Nothing sexual, no sign that he recognized me, no evidence that I was using the weights he wanted. There was just something about the cold intensity of his eyes. That's what a ghost would look like, I thought. Trying to be inconspicuous among the living, but unable to fake mortal eyes. Or a serial killer. I noted my own odd reaction and forgot about it.

I saw him again later, across the locker room, strolling too casually among the lockers with no gym bag. He was looking at me again. When I noticed him, he looked away and left quickly.

My own reaction felt weird enough that I mentioned it to my wife last night. I'm sure she'd testify on my behalf.

I'm not particularly sensitive to body language in some situations. I rarely notice when I've accidentally insulted someone, for example. But I'm pretty good at noticing the silent signals people send. When I owned a bar in my youth, I always recognized when the guy who sat down was mad at his boss and looking for a fight. I could usually predict which girl would leave with which guy. I trust my instincts about people. The guy at the gym broadcast unfamiliar signals, so I felt uncomfortable.

But I live my life mostly within the strict confines of science and business. This morning I returned to the mundane world of researching the history of iodine.

At the end of the 19th century, iodine was a miracle drug. If a doctor couldn't diagnose your ailment, he prescribed iodine and it worked more often than you'd think. Humans need iodine, modern diets tend to be deficient, science recognizes this. Iodine deficiency remains the number one cause of mental retardation in the world. Since the U.S. government mandated trace amounts of iodine be added to salt in the 1920s, the devastating diseases caused by deficiency have disappeared from the U.S. Many people believe we still don't get nearly the ideal amount. Even though Japanese people get fifty times as much as we do in their traditional diet, American doctors and scientists seem frightened by some vague threat posed by getting too much.

Something happened sometime in the early 20th century that made doctors suspicious of iodine and I hadn't been able to find out what that was. That's what I was researching this morning. Despite the health food store saleswoman's opinion of my brainpower, I wanted to give it a try but I didn't want to be foolish about it. Interestingly, the sternest warnings about the vague dangers of excess iodine seem to come from the drug companies that would rather sell you their own pills. But that's probably a coincidence. I found rumors of some scientific study that everyone believed, but that has since been disproved. I tried to track it down, as well as what the danger is supposed to be, exactly. I wanted to try increasing my own iodine levels, but I sure didn't want to give myself some disease.

Sometimes, taking too much iodine for a week or so can make the thyroid gland swell painfully. This is called "iodine thyroiditis." If you stop taking iodine, the condition goes away.

Two other diseases of the thyroid, "Graves Disease" and "Hashimoto's Disease," are autoimmune diseases. Graves causes the thyroid to over-produce hormones which speeds the metabolism; Hashimoto's has the opposite effect. Both diseases seem to have a genetic component. I don't think anyone believes iodine plays a role in Graves, but some people say that high doses of iodine have been "implicated" in Hashimoto's. Because it was described in 1912, maybe this was it. But the connection didn't seem conclusive enough to be the smoking gun I was looking for. Maybe a combination of possible dangers led to iodine's shady reputation.

I suspect that the incident with clioquinol in Japan entered the medical consciousness, including drug and insurance companies. Back in the 1950s, when people aspired to buy fancy $15,000 suburban homes, no doctor would forget a drug maker paying out nearly a billion dollars in claims for a drug that was basically just iodine. Professors would teach it to their medical students, pharmacy companies would remind their sales reps, insurance companies would include exceptions in their policies. I thought this case might be a big reason doctors still warn against taking too much iodine.

But I can't be the only person in the world who realized that Japanese people got so much iodine in their traditional diet that they had a special vulnerability to another massive dose. Surely there was more to their nervousness.

Your thyroid gland, which is crucial to efficient metabolism, is very sensitive to iodine. Without enough iodine, it can't do its job very well. Your metabolism slows, you tend to gain weight. When people don't get enough iodine for a long time, their thyroid gland works overtime to compensate. It gets bigger, it grows new cells. It becomes visible as a huge lump in the patient's neck called a "goiter."

Sometimes, when people who have been extremely deficient for a long time suddenly get a big dose of iodine, their thyroid goes into overdrive. It's used to working hard. When you provide the raw material it's been craving, it goes crazy, like a race car when it's brake is released. It pumps out hormones that speed up metabolism, including your heart muscles. That can be dangerous. You want to be careful not to go from long term extreme deficiency to excess overnight. But that just seems like common sense, not a reason to be fearful of raising your intake nearer the level considered normal in traditional coastal cultures.

I discovered one fascinating and weird thing. The noted psychic Edgar Cayce (1877-1945) became an enthusiastic and vocal proponent of iodine. Cayce was (and remains) controversial. While in a sort of trance, he talked about past lives, gave psychic readings, and diagnosed illnesses. Many books have been written about him and he continues to have followers. I just never happened to learn anything about him.

What was interesting for our purposes was this: He claimed to have channeled a process of passing electricity through iodine that detoxified and activated it. He wrote six hundred papers on the matter. A huge cult of people began buying the Atomiodine produced by his method. One guy made a fortune selling the stuff. He lost his fortune when Morton Salt started producing iodized salt in 1925.

I can certainly imagine scientists discrediting iodine for no other reason than that Edgar Cayce believed in it. I needed to do some research on this Cayce fellow, who I'd heard of but that's it. First stop was Wikipedia.

When the page opened, a picture of Edgar Cayce stared out at me and I did a double take. It was the guy from the gym yesterday. Same face, same hair, same body type.

Same eyes.

How would you feel if this guy was watching you at the gym?

(Edgar Cayce)

By now it probably goes without saying that, despite my encounters at the gym and at the health food grocery, I became more determined to find some nice iodine pills to try as a supplement. Using different words in an Internet search yielded quite different results. Searching for "iodine + diabetes," for example, gets you to a whole subculture of people convinced that iodine is the fountain of youth and the cure for everything. It leads to

135

many suppliers who are happy to sell iodine supplements in much larger doses than my grocery store sells.

But when I searched the phrase "iodine for sale" I stumbled onto the U.S. Department of Justice website. It turns out, pure iodine (in crystal form) is considered dangerous and the government regulates its sale. It's not so much that the fumes emitted by piles of pure iodine crystals at room temperature can irritate your skin, eyes or lungs, which they can. Let's let the Department of Justice explain in its own words:

"Methamphetamine producers use iodine crystals to produce hydriodic acid, the preferred reagent in the ephedrine/pseudoephedrine reduction method of d-methamphetamine production. A reagent is a chemical used in reactions to convert a precursor into a finished product. The reagent does not become part of the finished product. The regulation of hydriodic acid by the Drug Enforcement Administration (DEA) in 1993 rendered the chemical virtually unavailable in the United States.

"Hydriodic acid can be produced by combining iodine crystals with water and some form of phosphorus, including red phosphorus, hypophosphorous acid, or phosphorous acid. In the methamphetamine production process, iodine crystals may be used to prepare hydriodic acid in a separate step or may be introduced directly into the synthesis of the methamphetamine."

That's right. It turns out, iodine is a major player in the dark world of drug manufacture and smuggling. Anyone who buys it with the intent to manufacture illegal drugs could face up to ten years in prison and a $250,000 fine. Penalties are doubled for a second offense. This might explain some of iodine's reputation, although it seems odd that the Justice Department has gone out of its way to provide a useful recipe on its website.

I have to wonder if the intense-looking, wild-eyed Edgar Cayce knew anything about this use for iodine?

I had seen references to iodine pioneer Dr. John B. Stanbury so I got his book, *The Iodine Trail: Exploring Iodine Deficiency and its Prevention around the World.* Stanbury is the Indiana Jones of iodine. He spent decades traveling to exotic countries researching it, meeting dictators and natives, setting up field labs, flying in and out of primitive towns on tiny airplanes. His life would make a great movie. Based on his research, he founded the International Council for the Control of Iodine Deficiency Disorders (ICC-IDD) which works to eliminate iodine deficiency around the world.

The Iodine Trail is mostly about Stanbury's life and travels, but one intriguing sentence referred to Jean-Francois Coindet. That sentence made me track down information on him.

In 1820 and 1821, less than a decade after iodine was discovered, Coindet (of Geneva) published three papers that changed history. He guessed that iodine might be the active ingredient in seaweed that cured goiters. Now that it was easy to isolate it, he tried treating goiter patients with the chemical itself, rather than the seaweed. He started out giving patients doses totalling 75 mg. per day. That's a massive dose. The current RDA of iodine is 150 mcg, so his patients received 500 times the RDA every day for two weeks; after that he cut it back a bit. Treatments could continue for over two months. He treated over 150 patients with impressive positive results.

His findings caused such a stir that other people decided this iodine stuff was too good not to try. That year, by one estimate, over a thousand other people received doses of iodine, many in much less controlled environments. A number of those people complained of stomach ailments and worse. Some people died of heart attacks and their survivors blamed iodine.

A well known doctor named Colladon decided iodine was dangerous. Based on his own studies using massive doses of iodine on dogs he described iodine as a "corrosive poison." Of the nine humans he treated, six complained of stomach aches or discomfort. A well known socialite died of respiratory problems after two months of treatment with iodine. In the town of Geneva, there was much anger and outrage. The local doctors called a special meeting and voted that iodine was too dangerous to be used without a prescription. Rumors that Coindet was afraid to leave his home for fear of being stoned to death have been refuted. Still, it was probably not a happy time for him. He never again wrote about iodine and it fell out of favor for half a century.

In 1850, Frederic Rilliet also described the dangers of too much iodine. If someone who has been iodine deficient for a long time and has a goiter takes too much iodine, their thyroid goes into overdrive. The resulting flood of hormones can speed their metabolism to deadly levels. In 1910 Dr. Theodor Kocher gave that danger a name, the "Jod-Basedow phenomenon." Dr. Kocher had performed two thousand operations to remove enlarged thyroid glands and won a Nobel Prize for developing surgical techniques, so his disdain carried a lot of weight. Giving the condition a scary name probably didn't help iodine's reputation much.

It occurs to me that Dr. Kocher made his living by surgically removing thyroid glands; if swallowing tiny amounts of iodine could actually cure goiters, that would not be good for his business or legacy. I haven't noticed anyone else suspecting Dr. Kocher of having a conflict of interest, so let's blame the idea on my own sarcastic inner voice. Interestingly, this phenomenon was confirmed decades later and more scientifically by John Stanbury, the guy whose book got me started in this direction.

In 1948, two scientists from the University of California at Berkeley reported on a weird side effect of iodine that now bears their names, the "Wolff-Chaikoff effect." This might actually be the study I'd heard rumors about. According to this report, taking too much iodine can force your thyroid into a counter intuitive little dance that actually reduces its output, causes goiters, and slows metabolism. That is, a small amount is good and improves thyroid function, but a big dose can have the opposite effect.

This paper became well known and influential. The problem is, they did their experiments on rats, not humans, and it sounds like the effect has never been proven in humans. In fact, some people dispute the original results with the rats too. Here's where reading on the Internet gets interesting when you're not confident of your own science: The folks who write the clearest and most compelling descriptions of why this study was crazy are doctors who are, on the same website, trying to sell you iodine.

Today, as far as I can tell, iodine remains controversial in a couple of ways. Around the world, iodine deficiency remains a serious problem leading to goiter and mental retardation. In the U.S., severe deficiency is rare, but some people feel we aren't getting the optimum amount. No one seems quite sure what that amount is, but proponents seem to focus on 12mg per day, probably because this is about how much Japanese people got in their traditional diet with no ill effects. That's 80 times more than the RDA, but a fraction of what Coindet used. Other people remain concerned that too much iodine can be dangerous, at least for some people.

Some of this concern is undoubtably based on good science and medicine, but even the best science is probably still overcoming the bad reputation iodine got in the days of Coindet, then with the "Jod-Basedow phenomenon," and then in the days of Cayce, and then again with the clioquinol incident, then with the Wolff-Chaikoff paper. One could argue that each one of these involved special situations or flawed experiments, but getting really bad press every fifty years seems to have colored iodine's reputation over time. To say nothing of the reputation it's recently gotten by hanging out with drug dealers. And now, by having a guy like me write about it.

The interesting clues about iodine for our purposes as detectives are:

1) an iodine deficiency in a developing baby can cause neurological diseases,

2) The fungicide maneb interferes with the use of iodine in animals,

3) Maneb seems to act as an amplifier for some toxins. In combination with paraquat or rotenone, it causes the symptoms of Parkinson's Disease.

Those are the clues we might write down on three by five cards. Staring at those cards for a while, we wonder: could an iodine deficiency be a contributing factor in neurological diseases? That's just my own thought, I haven't heard it elsewhere. Even if someone else decides it's interesting, it would probably meet resistance from the scientific and medical community.

After you hear enough bad rumors, it doesn't matter what the truth is. If iodine was a teen-aged boy, your mother would not let you date him.

Superoxide

Superoxide is oxygen with an extra electron. It's a "free radical." Our bodies make it all the time and we also get it from our environment. Like most things our bodies produce, it is both useful and dangerous. Oxygen can rust your car; superoxide would rust it faster. It can also "rust" blood vessels and pretty much anything else it encounters. Its destructiveness makes it a powerful weapon to use against bacteria or other invaders. It can also react with metal ions, like copper or iron. The products of those marriages, like superoxide itself, are called "Reactive Oxygen Species" or ROS and they remain eager to react with their environment. Our bodies also make the antidote to superoxide, which is called "superoxide dismutase" or "SOD." There are three different types of SOD that contain various metal ions in their active sites, like copper, zinc, or manganese.

The big white blood cells that surround and ingest bacteria (phagocytes) produce superoxide to kill those bacteria. Superoxide is dangerous to the phagocytes, too, so they only make it right before they use it. The superoxide produced will react with itself or with SOD to form hydrogen peroxide and oxygen almost instantly. Interestingly, phagocytes can also produce hypochlorite, which is household bleach. Besides invading bacteria, phagocytes also surround and consume dead cells or dying cells, as well as other harmful things they encounter in the blood. Phagocytes are authorized to use superoxide. It's a good tool for them to have.

Mitochondria are little organelles within cells that help convert chemicals into energy. They produce superoxide all the time as a by-product of their work. The variety of ROS they produce seems most interesting to our puzzle.

We don't want superoxide roaming around our bodies and brains destroying things. Some things we eat contain "antioxidants" that destroy it. Things like vitamin C and vitamin E are antioxidants. That's one reason we're encouraged to consume them. We also produce SOD to roam our bodies and break superoxide down into less dangerous chemicals.

Here's what was interesting to me: People with an inherited form of ALS don't produce SOD very well. Their mutated SOD seems to work backwards, producing instead of destroying superoxide. Some people think this is a very important clue to understanding the disease. Equally interesting to me: mice that are genetically engineered to lack a gene for producing enzymes needed to make SOD are more sensitive to drugs like paraquat. That's because paraquat generates superoxide.

According to Nagendra Yadava and David G. Nicholls at the Buck Institute for Age Research (Novato, California, 2007), even tiny amounts of rotenone also cause the production of superoxide. As they say, "As little as 5 nM rotenone increased mitochondrial superoxide levels and potentiated glutamate-induced cytoplasmic (calcium) deregulation, the first irreversible stage of necrotic cell death."

Oxidative stress is what you call it when you have too much superoxide or ROS in your body. The role of superoxide and the various kinds of ROS has captured much attention in the scientific community.

In a March 2011 paper, Iman M. Mourad and Neveen A Noor of Cairo University reported that aspartame (an excitotoxin) causes oxidative stress in rat brains. The amount of this stress seemed to depend on the duration of the dosing. At first, not so much. After continued exposure for a few weeks, it became much more dramatic.

Rather than look at a fruit bat or cyanobacteria starting a chain of events, (that might end with a deficiency of SOD and therefore too much superoxide, which might kill nerve cells, for example) many people start at the other end of the process. They examine the nerve cell itself. If a variety of ROS or superoxide is causing the damage, maybe they can work their way back through the process to find a step they can correct.

Turning Nerve Cells On and Off

"Acetylcholine" is one of the chemicals our bodies produce to turn nerve cells on. Because it's a long, scary word, I'll call it the "on switch" although that might not always be accurate. "Acetylcholinase" (also known as acetylcholinesterase) destroys acetylcholine and turns a nerve off. I'll call it the "off switch." We need both the "on switch" and the "off switch" chemicals for our bodies to operate correctly. The balance between the two chemicals seems to be critical. Too much of either one and a person can become paralyzed or die.

If you don't have enough "on switch," you could either augment the "on switch" chemical (maybe you'd eat more lecithin, for example, because it contains choline, one of the raw materials your body uses to produce it). Or you might try to inhibit the "off switch." Either way, you wind up with more "on switch." The drugs prescribed for people with either ALS or Alzheimer's often inhibit the "off switch." They are called "acetylcholinase inhibitors."

But too much "off switch" is just as dangerous. Some deadly poisons work by inhibiting the off switch. Sarin nerve gas does this and so does the deadly botulism toxin.

B Vitamins

In 2008 a scientist at the University of California Irvine performed an experiment that stunned everyone who'd ever thought about Alzheimer's Disease. Working with mice and using only large doses of one particular B vitamin, he claims to have reversed the disease completely in a matter of months. This was not one of the flashy vitamins that celebrities tout on TV. This was one we've known about for a long time but has remained sitting on the sidelines, not interesting enough for publicity, with no significant cheering section.

By the time I read about this, I was not particularly surprised that a B vitamin had a dramatic effect on mice with a neurological disorder; these vitamins had been showing up with such remarkable frequency whether I was reading about fruit bats or bacteria or any of the several diseases I've gotten interested in. Many of them seem to work in cooperation with each other. Too much of one depletes one of the others, for example. Most seem to be involved with metabolism and energy production.

The B vitamins have such a colorful history they might merit their own TV series, each week featuring a different vitamin as hero. Many of them were discovered as cures for terrible diseases that killed and disfigured thousands of people. The discovery of each one changed the world, and it appears that we're not done learning about them. We could call our TV series "The B Team."

The first B vitamin was discovered by George Whipple in 1920. George was trying to figure out a cure for pernicious anemia. Patients with pernicious anemia don't make enough red blood cells. The disease was usually fatal and there wasn't much anyone could do for its victims. In what seems like brilliant thinking to me, George decided to make dogs anemic by removing some of their blood, (which effectively made them instantly anemic) then experimenting with foods to see which ones helped them make new red blood cells.

It turns out that feeding the dogs liver encouraged their bodies to replenish their red blood cells faster than anything else. Now he just had to figure out why.

The first thing he discovered was that liver is rich in iron. The iron stimulated red blood cell production. That's how we discovered the importance of dietary iron.

But the liver contained another substance that also stimulated red blood cell production. That substance is now known as vitamin B-12. A severe lack of B-12 causes neurological symptoms very much like the diseases we're talking about.

Vitamin B-12 contains the element cobalt. Interestingly, neither animals nor plants can manufacture this vitamin. Only bacteria and yeast can make it — the rest of us get it from them. Animals can get some B-12 from bacteria living in their intestines, but most animals don't get enough that way. If you're lucky enough to have more than one stomach (like cows do) then you probably house and feed such a rich colony of bacteria — and have enough intestinal real estate downstream from them to absorb it — that they will manufacture all the B-12 you need. Eat your grass, you'll be fine. Your body will store the surplus throughout your body, but mostly in your liver. Or, in the case of some animals, in their kidneys. These organs send all the rest of your cells what they need, as they need it.

Those of us with less glamorous alimentary canals need to eat something from time to time that contains B-12. This usually means eating some meat, but it could mean eating some bacteria or yeast. Vegetarian animals, like fruit bats, must snack on the occasional insect (or drink stagnant water containing bacteria) to get the B-12 they need to stay healthy.

I notice we're talking about fruit bats again. The B-12 requirements of fruit bats has actually been the focus of several studies. If you compare humans to fruit bats by body weight, it appears that fruit bats require a B-12 concentration between five and fifteen times greater than humans. They seem to store it in both their liver and kidneys. Interestingly, fruit bats in captivity begin to lose the concentration of B-12 almost immediately. We think that's because their captors are too fastidious: they never feed them an insect or give them dirty water, so the bats don't replenish their supplies. The longer a fruit bat lives in captivity, the less B-12 its body contains. If you feed a fruit bat nothing but clean fruit and clean water, at some point it will die from a lack of vitamin B-12.

Before that happens, a deficiency of B-12 causes neurological disorders in fruit bats and it seems to be worse if we give them folate (aka Vitamin B-9). Interestingly, folate might inhibit the body's ability to absorb zinc. One study refers to the "well documented interaction between folate and B-12" and describes the results as often being "crippling." Like I said before, the B vitamins interact, but not always in a helpful way. I bet we learn a whole bunch about B vitamins in the next few years and I bet some of it turns out to be quite important.

Thiamine, aka vitamin B-1, is another interesting case. Severe thiamine deficiency causes "beriberi," a disease with symptoms that include severe fatigue, cardiovascular problems, lack of muscle control, and an inflamed nervous system. A lesser deficiency causes irritability, depression (to the point of attempted suicide), fearfulness, agitation and memory loss. The term "beriberi"may have come from an ancient phrase meaning, "I can't, I can't."

There are several varieties of thiamine. Studies indicate that people with ALS have less of one variety in their brains than nonpatients. People with Alzheimer's also seem to be deficient. We need thiamine to produce ATP, which is vital for energy. Thiamine also seems to inhibit acetylcholinase, the enzyme that breaks down acetylcholine (at least it did in a study involving electric eels). Too much acetylcholinase (or maybe uninhibited acetylcholinase) leads to a lack of acetylcholine, the chemical our nerves use to communicate with each other. That is, a thiamine deficiency might encourage your "off switch" chemicals, which is the opposite of what ALS and Alzheimer's patients want happening. Interesting.

During the 1800s, beriberi was often fatal. On long sea voyages, it routinely killed a high percentage of the sailors (sometimes over half of a ship's crew). It was common in Asia. In 1883 a scientist named Takaki Kanehiro figured out that it had something to do with diet. That was a huge step. The disease had been killing people for many years and there wasn't any treatment. Gradually, people figured out that some "accessory factors" in the diet must be involved.

In the late 1800s, Christiaan Eijkman's started feeding his chickens polished rice and they got sick. When he switched back to brown rice, they recovered. So he did some experiments. He fed some chickens polished rice and fed others brown rice. He gradually figured out that the ones eating exclusively white rice were getting beriberi. Turns out, the brown outer layer of rice (the bran) contains the magic "accessory factor." Like chickens, people who only ate polished white rice were the ones getting the disease. Eijkman made the big connection but then he got malaria and others had to help put it all together over the next few years. In retrospect, it seems so obvious:

At around the time of the American Revolution, Japanese people decided that white rice (without the bran) was more appealing than brown rice (with the bran). Mills powered by waterfalls made it cheap to polish off the vitamin-rich brown layer, so white rice became popular. No one knew there was a health disadvantage to that, so that's what people ate for over a hundred years.

In 1929 Eijkman and Hopkins were awarded the Nobel Prize for coming up with an extract of rice bran to cure beriberi. That extract was thiamine. Later, when people discovered that it wasn't technically an "amine" they changed the name to thiamin without the final "e" but both spellings have survived and seem to be accepted. Today, beriberi is rare in advanced countries because so many of our foods are fortified with vitamins. However, people who abuse alcohol may come down with it because alcohol interferes with absorbing thiamine in the first place.

We've known that thiamine is involved with the metabolism of carbohydrates for a long time. In his Nov. 7, 2011 blog post, physician and noted thiamine expert Derrick Lonsdale described that discovery:

"In 1936 Sir Rudolph Peters did the research that led to our understanding of (thiamine's) vital actions. ... The study involved the biochemical activity of thiamine deficient pigeon brain cells compared with cells that had adequate thiamine. There was no difference in the activity of the thiamine adequate versus thiamine deficient cells until glucose was added in the experiment. It was immediately apparent that the thiamine deficient cells remained inactive whereas the thiamine sufficient cells became active....It must be more than obvious that our food is the equivalent of gasoline in a car, but it also requires chemical components that are the equivalent of the spark plugs and other components of adequate "engine" function." (from o2thesparkoflife.blogspot.com/ Used with permission)

Dr. Lonsdale has been a leading proponent of the idea that thiamine deficiency may play a role in autism.

You'd think that would be the end of the episode about thiamine for our TV series, but it has recently become worthy of a sequel. In 2007, English scientists led by Professor Paul Thornalley discovered that there was a flaw in the way we had always tested for thiamine in a person's body; in some cases this flaw meant doctors weren't detecting thiamine deficiencies. With a new test they discovered that diabetics have a dramatic deficiency in thiamine. Diabetics in the study had only one-fourth as much in their bodies as they should. Further studies suggested that it wasn't because of lack in their diet; rather, diabetics seemed to eliminate thiamine much too quickly.

This deficiency, some now believe, might be what causes the damage from the disease; the retinopathy (blindness), neuropathy (pain or lack of feeling), kidney damage, and probably the damage to the heart and blood vessels. Beriberi and diabetes share several symptoms. Both cause fatigue, both can cause pain, both can cause confusion and heart problems.

No one (that I know of) has made this claim, but it's almost like diabetics have their own brand of beriberi. Scientists are now investigating whether B-1 might help diabetics.

Thiamine comes in several varieties; most of these are water soluble. All the water soluble vitamins, like vitamin C, get processed through our bodies quickly and eliminated. We don't store them very well. Some nearly identical forms of thiamine are fat soluble and therefore don't get eliminated as quickly. They also pass the blood/brain barrier more easily. These have been sold as supplements (especially in Japan) for decades. People buy them to improve their alertness and focus, as well as to improve athletic performance. I bought a jar of one type (sulbutiamine) and have tried small amounts of it. I do think that I'm remembering facts better than I did a year ago and don't notice any negative effects. On the other hand, I've tried all sorts of vitamins and minerals during the course of writing this book so there's no good way to tell what's helped.

Knowing that the first form of fat soluble thiamine (allithiamine) was discovered in garlic, I've also increased my garlic intake. Luckily, I love garlic. Allithiamine's scientific name is the catchy "thiamine tetrahydrofurfuryl disulfide" (or TTFD). A synthetic form of allithiamine was developed in Japan by Takeda Chemical Industries in Osaka, Japan under the name

Alinamin. Scientists have been experimenting with this in the U.S. since the early 1970s but I don't believe the FDA has yet approved it for sale to regular citizens. In one study, it seemed to cause at least modest improvements in some Alzheimer's patients.

Vitamin B-2, aka riboflavin, is also involved in metabolism. It's the one that makes your B vitamin pill look yellow or orange. If you take a lot of it, your urine may look fluorescent yellow. There's got to be an episode in our series devoted to that.

One of the most intriguing B vitamins for our series is B-3. It is found in several forms, but you probably know it as niacin. Without enough niacin, people get sick with "pellagra." This disease features four major symptoms: diarrhea, dermatitis, dementia, and death. People identified pellagra as a specific disease in 1735, but didn't know what caused it. As recently as 1916, up to 100,000 people in the southern United States alone suffered from it. The team of scientists who figured out that niacin cured the disease were named Time Magazine's men of the year in 1938.

When you eat niacin, your body converts it into a different form of B-3 called niacinamide. Something in that process of conversion encourages your body to produce more HDL, the "good" cholesterol. Something in that process also tends to make your skin feel hot and get red, which is known as the "niacin flush." Once it's converted to niacinamide, it no longer has much effect on cholesterol and your body doesn't keep it around more than a few hours.

You can buy either kind of B-3, niacin or niacinamide. Niacin has become popular because of the cholesterol effect. Niacinamide is supposed to be good for your mood and complexion and has been used to treat acne for years, but that's about it. Acne is primarily an inflammatory response to minor bacteria; niacinamide is an anti-inflammatory. Interestingly, it can cross the blood/brain barrier so might also reduce inflammation within the brain.

Dr. William Kaufman believed in niacinamide the way some people believe in religion and did a lot of research on it beginning in the 1930s. He believed that a deficiency caused a whole series of mental problems from "mental fog" to "impaired memory" to just being in a bad mood all the time. He ate huge doses of it daily for 60 years with no apparent ill effects and seemed to be in a pretty good mood most of the time. He maintained the stuff was so benign you'd have to eat a pound of it every day before you started having problems. He's also the guy who discovered that your body processes it very quickly so if you want to maintain high levels in your blood you've got to take it several times a day. Dr. Kaufman still has some ardent followers who proclaim the virtues of niacinamide, but I haven't seen much science to back up most of their claims. At least not until the very recently.

Then, in 2008 Dr. Kim Green of the University of California at Irvine decided to try niacinamide on mice with Alzheimer's. His results, published in November of that year, were remarkable.

To study Alzheimer's we needed animals to study. At some point, someone (and I know there must be another interesting story in there) genetically modified mice so they always get Alzheimer's. Scientists treat these defective mice with different medicines and nutrients to see which ones benefit them. The question you're asking is the same one in my brain: if they can modify mice genetically so they always get the disease, you'd think they would understand exactly what change they were making. You'd think there would be a more direct way to reverse engineer the process. I can't answer that for you, because I don't know. Apparently, we have a strain of mice in labs everywhere that always develop Alzheimer's Disease.

Dr. Green took a bunch of these Alzheimer's mice and gave them high doses of niacinamide at regular intervals. A human equivalent dose would be between 1,000 and 2000 mg. three times a day. That's a lot, but one can buy 1000 mg. niacinamide pills at the health food store, so it's not completely unreasonable. And, obviously, that's a small fraction of the amount Dr. Kaufman thought might cause negative side effects. Dr. Green's results stunned the scientific community: after three months, the mice showed no symptoms of Alzheimer's. They were, apparently, completely cured. As long as they kept getting the B-3, they stayed well, But when the vitamins stopped, they began to regress.

Interestingly, the control group of normal mice that were fed the same vitamin regimen displayed signs of improved memory as well.

Of course, we can't get unduly optimistic. I haven't found a nice explanation for the chemical processes involved. But if the process that damages the brain in Alzheimer's, ALS, and Parkinson's is inflammation, and niacinamide is an anti-inflammatory, that may be the answer. We have many reasons to remain cautious until more test results are announced. Maybe mice that are genetically engineered to get Alzheimer's are a lot different from humans who get it naturally. Maybe niacinamide works different in mice than it does in humans. Maybe niacinamide will have a positive effect on humans, but maybe not. And maybe there was a flaw in the original experiment or the reporting of it, or my understanding of it. It is way, way too early to get overly optimistic. Still, it's interesting. They seemed to have cured mice of Alzheimer's by giving them regular, massive doses of niacinamide.

Less than two years after these findings were first published there are already several studies involving humans. That rush to clinical trials indicates that someone feels they're on to something valuable.

Before you start dosing yourself with niacinamide (which I did, of course) you should know that there is one negative side effect of either niacin or niacinamide: it can reduce your sensitivity to insulin. That means, taking big doses of either form of B-3 can raise your blood glucose levels. In diabetics, this is a dangerous disadvantage. Sometimes doctors forget to mention this to their diabetic heart patients who take niacin to reduce their cholesterol. If you're having trouble with your blood glucose levels and happen to be taking big doses of niacin for cholesterol, you should ask your doctor about it. Gently suggest that he might enjoy buying a copy of this book.

As far as the mystery we're studying, I wonder if the following are coincidences or clues:

ALS, Parkinson's and Alzheimer's all seem to include inflammation of nerves.

Acne involves an overaggressive inflammatory response to a trivial bacterial infection.

Clioquinol is a form of iodine that kills germs, especially on the skin. It's been used to treat acne. There have been several unscientific reports of it having a dramatic positive effect on Alzheimer's.

Minocycline is an antibiotic used to treat acne. It also has an anti-inflammatory property and has been widely used for over 20 years. There have been several unscientific reports of it having a dramatic positive effect on Alzheimer's.

Niacinamide has been used for decades to fight acne and improve memory and mood; one recent and very dramatic study claims it has reversed Alzheimer's in mice.

In February, 2012, researchers at Case Western Reserve University School of Medicine reported that the drug "bexarotene" began to eliminate the plaque found in the brains of mice with Alzheimer's within 72 hours. In that time, the mice's memory patterns also seemed to improve. Bexarotene is already approved to treat certain kinds of skin cancer and spot baldness. Although it's expensive, it's also used to treat acne.

The Forensic Approach

Sherlock Holmes would be proud of today's criminal detectives. Not only do they notice details, as he did, and look for clues that others miss, but they also employ science. They gather information that may or may not be relevant to solving the case. Sometimes this information just confuses and overwhelms the rest of us, but it's critical. As Sherlock himself said, "It is a capital mistake to theorise before one has data. Insensibly one begins to twist facts to suit theories instead of theories to suit facts."

Holmes did something else that elevated him above the common detectives in the literature of his time: he considered the "absence of something expected" as good a clue as the "presence of something unexpected." He solved one murder by noticing that the guard dog didn't bark when the intruder entered. This told Holmes that the dog must have known the killer. A dog that doesn't bark can be a great clue.

Some toxins cause symptoms in monkeys but don't affect fruit bats and rats. I'd love to know why that is, but have not been able to get an answer. Everyone I asked dismissed the question by saying "different animals just react differently." I remain unsatisfied. I think I hear a dog not barking.

In the mysterious case of our hypothetical homicide victim, back in the morgue, the medical examiner reconstructs a microscopic history of the murder. It's not enough to understand that the cause of death was a bullet wound. The M.E. wants to know the path of the bullet, what organs or vessels it injured, and how the body reacted. The gunshot wound, obvious though it might be, may not have been the actual cause of death. The subsequent bleeding or infection might have been fatal, but it's also possible the victim was hit by a truck as well a bullet. Which came first? The details of the process are as important as the motive of the murderer.

Neuroscientists do the same thing. They look for clues within victims and within cells and in microscopic traces of chemicals.

The "bullet" may be a toxin or some other environmental factor, but these particular scientists aren't looking for anything that big and obvious. The "victim's" reaction to that factor may be what actually does the damage. The cure could involve identifying and eliminating the toxin (for example) or it could mean changing the body's reaction to it. A few processes within human bodies intrigue most of the people who study these diseases.

One that shows up time after time is called "inflammation."

Inflammation

Our bodies respond to threats effectively and aggressively. But sometimes, the response is too aggressive and causes more damage than the original threat. We kill the mosquito so forcefully we leave a bruise on our own arm.

When a spider bites us, or we get a sliver, or stub our toe, our bodies launch a complex defense strategy called inflammation. There is growing suspicion — bordering on conviction — that an overly aggressive inflammatory response is what actually does the damage in ALS, Alzheimer's, MS, Parkinson's Disease, and other neurological diseases.

Inflammation is a complex dance of chemical stimulation and response. Some things signal our bodies to do one thing and others stop them. The initial trigger, perhaps the toxin from a spider bite, is like a stop light changing from red to green. The body responds with one chemical stepping on the gas, another shifting gears, a third turning the steering wheel, and yet another ready to step on the brake if a pedestrian steps onto the street. When everything works correctly, the passenger (our conscious brain) doesn't even notice that anything happened. It continues talking on its cell phone while our bodies drive on down the road. But if one function is out of whack, chaos ensues. If the brakes don't work or the gas pedal sticks or the steering wheel becomes loose then we find ourselves in a hospital room getting a tune up.

When you get a sliver in your finger, the first responders are the communications team. The cells near the sliver release chemicals called cyto-

151

kines. Cytokines are the body's alarm bells: they alert other cells to the threat. One cytokine is called bradykinin. Bradykinin communicates with your defensive team automatically, even while you're sleeping or unconscious. It also communicates with your brain, alerting your conscious mind to the problem. It does this by making the area near the sliver more sensitive to pain. You won't forget you've got a sliver in your finger thanks to bradykinin. Maybe you'll pull it out before it gets infected.

Bradykinin makes the blood vessels expand near the sliver so that more blood reaches the problem. The area around the sliver looks red and swollen because of this. That's where the word "inflammation" comes from: the red fiery look of the skin around an injury. Bradykinin is what causes your blood vessels to dilate.

The blood vessels don't merely swell. Bradykinin changes the permeability of their surfaces as well, allowing white blood cells (leukocytes) to pass through the vessel's membrane to get out there and fight the danger. Some leukocytes become huge macrophages; these engulf dangerous particles and also tell their cousins the lymphocytes to join the battle.

Once your body feels the threat is gone, it begins to reverse the inflammatory response. If the threat never goes away, your body continues to fight it but the campaign isn't always tightly focused on the location of the threat. Gum infections are notoriously hard for your body to overcome. They fuel a heightened inflammatory response throughout your body that can last for years. Fat cells trigger an inflammatory response through your body. Being overweight or having gum problems means your blood vessels are under constant stress, as if fighting bee stings all over your body. You're a lot likelier to have heart and blood vessel problems as well as other health issues that relate to inflammation.

The pain and swelling of inflammation can be useful to solving the problem of a sliver in your finger. But once we've removed the sliver and cleaned the injury, we don't want continued pain and swelling. We might put some ice on the site because cold tends to make nerves less sensitive to pain and make blood vessels shrink. Or, we might take an aspirin or other pill to interrupt the inflammatory response. Different medicines stop the inflammation at different steps in its process.

If you inject the part of a mouse brain called the hippocampus with a tiny amount of bradykinin, within days the mouse shows the symptoms of Alzheimer's.

Another cytokine is "tumor necrosis factor" or TNF. TNF is manu-factured by macrophages, large cells that fight infection and other assaults. It triggers other cells' defense procedures as part of inflammation. Those cells have a specific receptor site for TNF. Some people have a mutation in the gene that creates this receptor site. This mutation leads to a disease originally called Hibernian Fever because it was first discovered in a kid of Scottish and Irish descent in 1982. The disease is a bit like the flu — fever, rash, puffy eyes, abdominal pain, and aching muscles. One drug that seems to work against Hibernian Flu is called etanercept. It works to counteract inflammation caused by TNF. Remember the name of that drug; we'll be coming back to it.

The inflammation of rheumatoid arthritis is caused by TNF, so it has been studied extensively. Unusually high levels of TNF-alpha have been documented in patients with Alzheimer's so, in the last few years, scientists have begun trying to understanding its role in that disease as well.

Many scientists now believe that an overly aggressive inflammatory response is responsible for the cell damage of Alzheimer's, ALS, and the others. There is some evidence that anti-inflammatory drugs can ease at least some of the symptoms of these diseases. Ibuprofen seems to slow the progress of Parkinson's disease. Aspirin seems to reduce the inflammation in blood vessels that leads to coronary problems. And, in each case where a startling reversal of Alzheimer's symptoms was reported after giving a patient a drug for some unrelated problem, the drug in question had an anti-inflammatory property.

Research about the role of cytokines and inflammation in these dis-eases has become more intense in the last twenty years or so. It began with a concept but has already led to treatment experiments. Consider the history of the idea:

In 1989, Sue Griffin, Ph.D., published a study describing the associa-tion of Alzheimer's with a cytokine over-expression in the brain. Her re-search opened the door to this line of thinking. In 2007, Edward Tobinick M.D. published the results of a study that opened many minds to this idea. Dr. Griffin contributed extensive notes to the report. The idea was this: if TNF is elevated in the brains of Alzheimer's patients and TNF induces in-flammation, what would happen if we injected an anti-inflammatory drug specifically aimed at TNF into their nervous systems? In this study, they injected the drug etanercept (trade name "Enbrel") directly into the spine of Alzheimer's patients.

The patients showed improvement within minutes. This line of inquiry became one of the dozen or so hottest areas in the field beginning immediately following that study in 2007. Less than 20 years from concept to a new field of study.

Interestingly, niacinamide also reduces TNF in the brain.

Perhaps the least likely anti-inflammatory of all is a component of sewer gas. Our bodies produce hydrogen sulfide in tiny amounts and our cells use it to communicate with each other. Each molecule usually only lasts for a brief moment within our bodies. Some of the most intriguing new science I stumbled across deals with this smelly gas.

It feels like science is zeroing in on some dramatic answers. But let me step away from "science" for just a moment and let my imagination take over. The next section is just a weird daydream, not anything justified by scientific studies. When you get enough facts in your brain, sometimes you combine them in odd ways. Every now and then one of those daydreams proves useful, so we should all encourage them. We just can't take them very seriously.

Science Fiction

One entire group of scientists think diseases like Alzheimer's, ALS, and the others are caused by a "pathogen." That is, by some virus, fungus or bacteria that we just haven't discovered. History is littered with examples of diseases that remained completely mysterious to man until we discovered a microscopic critter behind the scenes. Diseases like smallpox, malaria, yellow fever, measles, mumps, plague, ulcers, some forms of cancer, and dozens of others fit this category. No one thought a pathogen caused them until someone proved that it did. People discovered the benefits of vaccinating against smallpox way back in the early 1700s. They didn't make the connection that tiny critters caused the disease until the 1880s.

If you deal with the elderly a lot, you already know that urinary tract infections, caused by bacteria, routinely make their victims confused and forgetful. Their children often think their parents have come down with Alzheimer's or some other form of dementia. The symptoms arise suddenly and usually disappear within days of treatment with antibiotics. If common forms of bacteria two feet from the brain can induce symptoms of dementia, is it such a leap to believe that some sort of "infection" causes these diseases all the time?

So far, we haven't discovered any microscopic culprit, so many (probably most) scientists don't subscribe to this idea. But the concept got me daydreaming and I had one goofy idea. I have absolutely no reason to believe it's true and don't think it probably is. On the other hand, I haven't seen any evidence that it's false either, so I'll toss it out there.

If a pathogen is involved, I wonder if it might not be cyanobacteria? If a colony managed to get into a person, perhaps because the human inhaled some pond scum water while water skiing or whatever, might these creatures that have proven so adaptable for billions of years, figure out a way to survive within, say, human nasal cavities? Our bodies may not produce an antibody for them, or cause a fever, which is about the only ways we know we're infected. If so, maybe they gradually build up in population for years, completely undetected, slowly releasing toxins. Or maybe keeping the toxins inside their tiny bodies until they all die for some reason. Our bodies process these toxins by using GST and similar enzymes. Until their numbers reach some critical mass, our bodies can keep up with the toxin load.

But then, if we encounter an additional stress — a romantic break up, a broken leg, whatever — our bodies release all the GST to fight the stress. Suddenly, we don't have enough resources to fight off the toxins, which trigger the whole inflammatory response. Depending on the toxin and our own reaction to it, we come down with Alzheimer's, ALS, Parkinson's or MS. Wouldn't that explain everything pretty neatly?

The only "evidence" for such a weird theory is that the chemicals that have caused remission (especially of Alzheimer's) in some patients happen to be drugs that fight inflammation and also inhibit exactly the kind of primitive bacteria we're talking about.

In 2010, Pamela Silver of Harvard Medical School injected some zebrafish embryos with fluorescently tagged cyanobacteria. When the fish grew up, they had glowing spots. The cyanobacteria were living and glowing within the fishes' tissues without causing any sort of defensive reaction at all. Her idea was to see if ultimately cyanobacteria could photosynthesize within the fish and contribute to its energy needs. Fish that produce their own food from sunlight would save fish farms a lot of money. There is, at this moment at least, a video online of these nearly transparent fish swimming around with glowing spots of cyanobacteria living within them. One can only hope that none of the cyanobacteria that produce toxins ever get the idea to colonize us in that way. And that none of them already have.

What if there really are colonies of cyanobacteria living within us that can, under the right circumstances, multiply and release toxins? How might we know and what would we do? We know that even the cyanobacteria that *can* produce toxins don't always do so. They could just live inside us for decades until something triggers the toxin-producing process within them. Even those that produce toxins don't necessarily release the toxins into the environment until something kills them. If our bodies don't notice them, we wouldn't produce antibodies or a fever.

Granted, this is a crazy, science fiction kind of idea. But there might be one way to test it: expose the critters to lots of oxygen. Most cyanobacteria hate oxygen and, in sufficient concentrations, it can kill them. I wondered if anyone had tried hyperbaric oxygen chambers on patients with these diseases?

Turns out they have. In mice, hyperbaric oxygen has had good results. In humans, it has had mixed anecdotal success. I don't know how much oxygen it would take or for how long or how often to be dangerous to anaerobic bacteria lodged deep within human tissues, but it might be something to think about.

Let me repeat that I don't actually have any good reason to suspect cyanobacteria are living in my nose right now. I don't actually believe it. It's just one of those ideas a person gets when he brainstorms with himself. Maybe you or I will actually brainstorm something that turns out to be useful. The fact that we'll probably be wrong virtually every time isn't a good reason to stop thinking creatively.

Remembering Vince

I've been thinking about my uncle Vince, a strong athletic high school senior struck down by polio. He spent weeks in a hospital, unable to move, barely able to breathe, slowly dying. The hospital had two iron lungs, massive, noisy machines that looked like space capsules from some ancient science fiction movie. An iron lung replaced the diaphragm muscle, inhaling for a patient when he couldn't on his own. Vince needed one, but both of these devices were occupied by other patients. There was no time to order another one, and Vince was deteriorating. His prognosis was grim.

Then the patient in one iron lung died. Sad news, certainly, but that made the machine available for Vince. He lived within that confined space for weeks, but gradually got strong enough to breathe on his own again. He could never again sit up without a brace, but at least he could breathe. Not by using his diaphragm like you and I do because polio paralyzed that, but with his few surviving chest muscles like we do when the doctor says "take a deep breath."

I knew Vince for over half a century after that, but it never occurred to me to ask him some obvious questions. What had he wanted to be when he grew up, for example? From the way his eyes lit up when describing the machinery of farming, I'm guessing he wanted to raise wheat in Kansas. Polio changed his options. He had a long time in that iron lung to think about his life.

He graduated from high school and enrolled in a small, prestigious college near his home. He studied biology. From his books, I know he also studied trigonometry and history, especially Civil War history. He read the Great Books. His buddies wheeled him to class in his wheelchair and pushed him home at night. He did fine in college, but the third year classes were on the third floor of the college building. There was no elevator. He didn't want to make his buddies carry him in his chair up those long wooden stairs, so he dropped out.

Then he learned to repair radios, clocks, and watches. Ultimately, he opened his own shop. Just when he was starting to get ahead, a partner stole all his inventory and sold it, keeping the money. Besides being paralyzed, Vince was now also completely broke and deep in debt. He just kept plugging along and, a few years later, paid off all the creditors. He even taught others how to make gears and bearings from flat scraps of metal, using precision lathes and other delicate machinery. He joined service clubs so he could "give back" to the community in the ways he could. He sang bass in a barbershop quartet.

Between all that, he taught me how to play chess, argued politics, and explained football to me. He subscribed to Scientific American, which meant that I got to read any article I could understand. Careful with money, he only invested in books and tools and never had much money for either one. So it surprised me when he gave me his prized set of Encyclopedia Britannica. He said it was just too difficult for him, he wanted a simpler encyclopedia. It's out of date by decades now, but I still own it and read a few pages from time to time. I also have his set of Great Books and have read many of them.

During those long days and nights in the iron lung, I imagine Vince deciding to learn what he could about the disease that changed his life. I bet he figured there was a chance he could nudge the science forward by taking the small steps within his power. He never said anything about it, but he started educating himself about biology, using the best tools of his day— college, then magazines and books.

Vince's little sister Katherine was probably inspired by his polio, too. After she graduated from high school, she got her nursing degree and spent the rest of her life helping sick people. When their parents died, Vince and Katherine lived alone in the family home. When a smart and charming soldier returned from World War II and proposed to Katherine, the one condition of her saying "yes" was that Vince could live with her and her new husband. The soldier agreed. Katherine was my mother.

Sometimes, we repeat the journeys that our ancestors took, but often with a twist. We return to the land our family once owned only this time, instead of taking a covered wagon, we fly. If our parent preached in front of a congregation, maybe we testify from a stage in Las Vegas. If our grandfather built fiddles, maybe we build songs.

I had not thought of this book as a journey like that. But maybe it is.

Back to the Bats

My brain keeps returning to this question: Why don't the fruit bats get sick? Although many things may cause or contribute to ALS (as well as Parkinson's, Alzheimer's, MS, etc.) the scientists studying lytico-bodig on Guam make a convincing argument that BMAA in cycad nuts certainly *can* cause the disease, at least some of the time. Fruit bats on Guam eat the cycad nuts and develop high concentrations of BMAA in their bodies. So why don't the fruit bats themselves get sick?

I e-mailed a couple of leading bat experts looking for clues. My thinking went like this: if ALS and Alzheimer's are both overaggressive inflammatory responses to a toxin, how is a bat's inflammatory response different than a human's? Perhaps I didn't make my question clear, but the answer I kept getting was: "nobody knows."

So I read a couple of college textbooks about bat physiology. Bats are obviously cool, but books about bats' biological processes are "less cool." One I read was probably a lot perkier in the original German. I wished I'd taken all the prerequisite courses first so I understood the language without a dictionary. Nothing I stumbled across seemed directly useful to our current conversation, but I did learn a lot and it seems a shame to waste all that study. In no particular order, here are my notes about bats. Memorize them because one might spark a conversation the next time you're chatting with a bat expert at a party:

Bats can swim. I never thought about that before, but apparently they can.

Vampire bats can walk around quite well. They can even leap from the ground and take flight, unlike most bats. It's a lot easier for most bats to release their hold on the ceiling of a cave, for example, and use gravity to help propel them for the first few seconds.

Fruit bats "commute" to food, often flying quite a distance. They are not fast fliers, but are strong fliers and can carry their young for several weeks after birth. We think they use smell to help them find food. Scientists believe that all bats can smell pretty well — at least as well as a human. The floral scent of eight plants that bats love each contained sulfuric compounds and fatty acid derivatives with a mushroom-like odor. These two elements are not found in the scents of related plants that are not pollinated by bats. Sulfuric compounds ... weren't we just talking about sulfuric compounds?

Relative to their size, bats have the largest and most muscular heart of all mammals. Their heart is more densely supplied with capillaries than any mammal. Unlike most mammals, most of the energy reserves in the heart are in the form of the chemical ATP.

Although not all bats can hibernate, those that do have some interesting features. Their heart muscle can contract and pump blood when very cold, even at temperatures near freezing. If you wake one from hibernation, it risks going into fibrillation because its cold heart muscles can't keep up with the speed of awakened nerves. We don't know much about how they regulate hibernation, but we think two chemicals may be involved: "vasoactive intestinal polypeptide" and "hibernation-inducing trigger." Based on its name alone, I'd bet on that second one.

During hibernation, a bat's blood sugar falls from 155 mg./dl (when active) to 28 mg./dl during hibernation. While hibernating, waste is not generated or excreted. In the lab, bats can sleep continuously for up to 200 days. A slowing heart rate is the first step we know of in hibernation. Bats can wake from hibernation over the course of a half hour, which is much faster than other hibernating animals. Although low temperature tends to slow metabolism, that's not all that's going on. Bats do something to actively suppress their metabolism. I just couldn't figure out what it was and I'm not positive anyone completely understands it. Hibernation is called "torpor" and a few kinds of bats can't enter torpor.

Here's a quote from *Biology of Bats:* "Bats have a powerful cholinergic innervation of the ventricles of the heart, which is uncommon in mammals." I believe this means that they have lots of nerves in their heart that are stimulated by acetylcholine, rather than the kinds of nerves that are stimulated by other chemicals.

Bats have "anastomoses" in their wings, which are shunts between arteries and veins that open and close at different times, allowing blood to get to capillaries or to bypass them. No one completely understands the function of these as far as I can tell.

Similarly, they have "venous hearts." These are not really hearts the way we think of them. Most animals, including humans, rely on the contraction of their muscles to squeeze blood back up to their hearts. The heart pumps blood through arteries down to our legs, for example. But the force of the heart is dissipated as the blood travels through hundreds of tiny capillaries and isn't strong enough to keep it moving out the other side. The blood returns to the heart via veins which have one-way valves in them. Every time you move your leg, your leg muscles squeeze the veins intertwined with them and the blood gets pushed along. If you never use your leg muscles, the blood tends to pool and your legs swell up. In their wings, bat veins are actually surrounded by muscles that constrict rhythmically.

That is, rather than rely on body movement, the veins squeeze themselves. This counteracts the centrifugal force of flapping that would normally prevent blood from returning to the heart. It's so effective that scientists refer to these paristaltic veins as "venous hearts." The veins sense when they're full of blood and squeeze. If provided with cow blood, the veins can keep contracting rhythmically for over a week after being removed from the bat.

Flying is hard work and requires both fuel and oxygen. Bats have smaller red blood cells than other mammals, but more are concentrated in each drop of blood. Their blood can absorb 25 percent to 30 percent oxygen. By comparison, small ground dwelling animals like mice can't absorb more than about 18 percent.

Bats can absorb some of the oxygen they need through their wings, which are basically two thin membranes separated by blood vessels. They can also eliminate some carbon dioxide through them.

Bats can't sweat. Now there's a fact you can surely work into a conversation. But overheating is deadly to them. Some bats lick their fur and then fan their wings to cool off by evaporation. Surely you'll want to impress a date with that image at some point.

This was fascinating to me, because it involved a potential toxin: Bats tolerate high concentrations of ammonia in their environment. In some cases, they live in caves with a concentration ten times greater than the concentration that would kill a human in one hour. I could not tell if anyone knows why this is or not.

Some bats can live over thirty years; fruit bats can live more than twenty. Most animals their size live maybe three years. "Baudry et al" suggest that aging is related to the "calpain" content of neurons. Calpain is a calcium dependent protolytic enzyme. No, that won't be on the quiz. One of its functions is to break down cytoskeletal proteins. Here's what was interesting: Calpain activity in bat brains is five to seven times LESS than in mice. Reptiles with long lives also have low calpain activity. The author of the book suggested that maybe someone ought to study this. I'm not sure there's any money to be made in studying an enzyme that affects the speed of aging by ten times, but I suppose he could be onto something here. In doing some casual research about calpain, I learned that it is a family of enzymes discovered in 1964. Not too much is known about it, but it's associated with calcium in the body. Overactive calpain has been noticed in patients with Alzheimer's and in people with cataracts. It is implicated in the degeneration of heart tissue after a heart attack and in the degeneration of brain tissue after a stroke or head injury. Interestingly, bats have much less active calpain than other creatures their size. I won't pretend I really understand all that, so don't worry if you don't either.

Fruit bats' kidneys don't concentrate urine as much as other bats' do.

Bats have high concentrations of urea in their blood. In one experiment, scientists showed that insect eating bats have concentrations four to five times that of omnivores. Fruit bats have higher concentrations of hydrochloric acid in their stomachs than other animals do. It may sterilize the fruit juice they consume (they squeeze out the juice with their tongues, swallow that, and spit out the pulp), but we don't really know.

Because food passes through them so quickly, bats may have increased enzymatic activity in their intestines. According to *Biology of Bats*, as of the year 2000 (the year the book came out) no systematic study of digestion in bats had been conducted. As to the enzymatic activity it said, "but little is known in this regard."

The only part of the bat brain that's smaller than a comparable sized insectivore (like a shrew) is the olfactory bulb. On the other hand, their hippocampus and amygdala are unusually large. The hippocampus is involved with memory.

Are there any clues in there? I can't tell. Maybe one of these bat facts answers some completely different question you or I will be asking next year. But, after reading through all this, I remembered another little tidbit I read months ago. At that point, it seemed even less relevant than venous hearts and bats going for a swim so I didn't include it. Now it seems quite interesting. Ammonia isn't the only toxic gas that bats endure much better than humans do. They also tolerate much higher concentrations of hydrogen sulfide.

The caves that bats love often contain high concentrations of hydrogen sulfide. In many cases, that chemical was part of the process that carved the cave out of the rock in the first place. Carlsbad Caverns can have high levels of hydrogen sulfide and is also home to millions of bats. One of the most interesting caves is Villa Luz cave near Tabasco, Mexico. Streams running through it look milky white from all the sulfur they contain. Humans take special precautions when exploring so they don't die from the hydrogen sulfide gas in the air. But bats thrive.

Bats have lived around hydrogen sulfide for a long time. Messel pit in Germany is one of the neatest sources of fossils in the world. It was once the bottom of a very deep lake formed by a volcano that routinely spouted poisonous gases. Because of its unique sediments, surviving impressions of ancient animals include detailed records of fine soft features like fur and wings. The most prevalent remains are from bats, nicely preserved from 48 million years ago. Scientists wonder what killed such perfect specimens. They wonder if it was the poisonous hydrogen sulfide gas from the volcanic fissures, or did the bats get trapped in the algae mats on the lake's surface?

Imagine that we were writing clues on three by five cards. Let's put a few of those cards down on the table at the same time for a moment and think about them:

Alzheimer's patients have too little hydrogen sulfide in their brains. People with Down's Syndrome have too much. Apparently, we need a fine balance.

In "type one" diabetes (sometimes called juvenile diabetes), the patient's body can't make insulin. Although we don't understand what "causes" type 1 diabetes, hydrogen sulfide is the final weapon that kills pancreatic cells that were supposed to produce insulin. Alzheimer's seem related to "type two" diabetes, in which the body becomes insensitive to insulin and may not manufacture enough of it. I'm not sure there's any connection between Alzheimer's and type 1 diabetes, but the fact that there is any relationship between hydrogen sulfide and any form of diabetes might be an interesting clue.

ALS, Alzheimer's, and Parkinson's all seem to involve excessive inflammation in nerve cells. Hydrogen sulfide can reduce inflammation.

Toxins like BMAA that cause the symptoms of ALS in humans don't seem to in fruit bats.

Bats process hydrogen sulfide differently than humans do; they can survive much higher concentrations.

If I were a scientist with a government grant and a nice laboratory, I might consider investigating the inflammation response of fruit bats and also how hydrogen sulfide works in their bodies. Instead, I e-mailed several bat scientists, explained what sort of book I was writing, and asked for information on the inflammatory response in bats. Each one answered the same question from the perspective of their own specialty.

A guy who specializes in bat immunity explained that bats survive some viruses and bacteria that kill humans. Some survive rabies, for example, while almost no humans do without quick treatment. Therefore, he said, we know that the bat immune system works somewhat differently than the human one does. But he said nothing about inflammation.

Another expert who had worked on Guam for years focused on why he disagreed with Paul Cox's theory, (he doesn't think Cox has rigorously proved his concept and disagrees with some of his premises about bats on Guam) but didn't really address the inflammation question.

Ecologists answer questions by turning to ecology, mathematicians draw equations on napkins, chemists haul out their test tubes and spectrometers. The main thing I learned from all this is that everyone specializes these days, but when you want an expert who has actually studied something very specific, it isn't always easy to track one down.

Fruit bat.
Photo by Adrian Pingstone

Hydrogen Sulfide

You can see why a guy like me would get intrigued by hydrogen sulfide. I'm still, at heart, a twelve year old boy who thinks weird smells are cool and hydrogen sulfide is the smell of feedlots, rotten eggs, volcanoes, and flatulence. Bats tolerate a lot more of it than humans do, which interests me because I'm looking for ways bats and humans differ. It's deadly to most cyanobacteria, but some kinds have learned to consume it. In fact, some thrive where oxygen-rich water adjoins a hydrogen sulfide-rich layer. The cyanobacteria use one layer during the day and the other at night.

By now you've figured out that whenever I hear a noise behind me I expect to see a giant blob of blue-green algae sneaking up on me. The fact that there's any relation between H_2S and cyanobacteria would be enough. But there's more.

As I said before, people with Alzheimer's or diabetes typically have too little hydrogen sulfide in their brains. Hydrogen sulfide is what kills pancreas cells and induces type 1 diabetes. But people with Down's Syndrome have too much. There's good evidence that neurological diseases often involve inflammation; hydrogen sulfide, in small doses, is an anti-inflammatory. But too much kills you. It binds to the same cell sites as oxygen does. In fact, there are some cases of ALS in people who had been exposed to dangerous levels of hydrogen sulfide and scientists wonder if it might have been involved. It has recently been declared one of the three gases that our bodies use as communicators, much as we use cytokines. (The other two are nitric oxide and carbon monoxide.)

That seems like a lot of clues to investigate. The body must have a whole process to manufacture hydrogen sulfide, and to regulate it, and to dispose of the excess. If something gets screwed up in that system, maybe we don't manufacture enough to counter inflammatory signals. Or maybe we get rid of it too fast. I decided I wanted to learn a little more about this smelly gas.

Energy companies remove hydrogen sulfide from "sour natural gas" by a "sweetening" process to make it less poisonous and corrosive. Then they add some "methyl mercaptan" to natural gas, which also smells like rotten eggs (and which, like hydrogen sulfide, is a component of flatulence, occurs in our brains, and is deadly in high doses) so we'll notice the smell of a gas leak.

Hydrogen sulfide is a very common chemical. In large doses, it's poisonous and sometimes used in suicides. In World War I it was used as a chemical weapon.

Our bodies make small amounts of hydrogen sulfide all the time. It relaxes our "smooth muscles," dilates and relaxes blood vessels, and increases the response of a particular kind of receptor in our brains that is involved with memory. Low hydrogen sulfide seems to play a role in both obesity and type 2 diabetes. In one study, diabetics had a fourth as much hydrogen sulfide as nondiabetics. The amount seems somehow related to how much fat a person carries around their waist.

Based on those facts alone, I would think we'd have entire universities devoted to studying this bad smelling gas. But no. Once again, it's too common, it's not patentable, it's not glamorous.

In 2008, some scientists at Harvard Medical School (including Frank Sellke, MD) published the results of a study about hydrogen sulfide's effect on inflammation. These scientists discovered that hydrogen sulfide reduces inflammation in pig hearts after an induced heart attack. It prevented the terrible damage to the heart muscle that normally follows any tissue being deprived of oxygen. That damage comes from the massive inflammatory response such an event triggers. Let them explain it to you in their own words:

"Conclusions: Therapeutic sulfide provides protection in response to ischemia/re-perfusion injury, improving myocardial function, reducing infarct size, and improving coronary microvascular reactivity, potentially through its anti-inflammatory properties. Exogenous sulfide may have therapeutic utility in clinical settings in which ischemia/re-perfusion injury is encountered."

I don't know how I could say that any clearer. (Being deprived of oxygen is called "ischemia." Infarct means cell death. Exogenous means something produced outside a body, while endogenous means something a body produces itself. Re-perfusion is bringing blood back to an area after it's been without it. Myocardial refers to the middle layer of muscles in the heart.) If you're going to do your own online research, get used to reading sentences like that. Scientists know all those words; they provide an efficient shorthand for them.

Even earlier, in 2005, researchers subjected mice to hydrogen sulfide. The mice entered a state very much like hibernation: their body temperature dropped dramatically, their breathing slowed from 120 breaths per minute to 10. It was almost like they became cold blooded animals whose temperature went down with the temperature around them and their metabolism slowed accordingly. They suffered no ill effects after being in this condition for several hours. This is interesting and surprising because mice, unlike bats, don't hibernate. In 2008, scientists used hydrogen sulfide to induce a kind of low temperature hibernation in rats in which they had induced strokes. The procedure reduced the amount of damage to the rat brains from these strokes.

Since then, a similar experiment using pigs and sheep failed, but experimentation continues. When you hurt your knee, you might put an ice pack on it to reduce the excess inflammation. Hydrogen sulfide seems to do something very similar to a brain. At least a mouse or rat brain. During hibernation, the body temperature goes way down. And so does the inflammation.

In another report, scientists from Harvard Medical School reported that inhaled hydrogen sulfide prevented nerve degeneration in mice with Parkinson's Disease (induced by giving them MPTP).

Mark Roth of Seattle, a guy smart enough to have received a Macarthur Foundation "Genius Grant," believes that we will soon be able to use hydrogen sulfide to induce a sort of temporary hibernation in humans, reducing their need for oxygen and reducing the damage from inflammation after various traumas including stroke and heart attack. He gave a talk to the TED forum (www.ted.com - then search "Mark Roth") about this in 2010 in which he claims to have completed the first phase of human trials (the safety trials) and expects to complete efficacy trials within the next two years. His company, Ikaria, hopes this leads to a profitable addition to their sale of nitric oxide.

At this point, you don't get to read a half a page that I wrote about the production of hydrogen sulfide. I double checked my sources and couldn't confirm what I'd written, so I deleted it. But it led me to an even more interesting place.

What I thought I'd read is that thiamine (vitamin B-1) was involved with the manufacture of hydrogen sulfide. Our bodies have a little production line in which we take homocysteine and convert that to the amino acid cysteine which we then convert to hydrogen sulfide. I could have sworn that I read, in a very official study, about thiamine facilitating that last transaction. It's possible that it does, but I can't find it again, so I don't want to tell you it's true.

But that impression led me back to thiamine, thinking about it in the context of diabetes and hydrogen sulfide.

Thiamine

As I tried to learn about hydrogen sulfide, I got distracted by thiamine all over again. I learned a whole truckload of information about this mysterious stuff. Believe it or not, thiamine turned out to be the most interesting part of this whole project for me. Yes, more interesting even than bats and maybe more important.

A lot of the material was written in language that I found nearly impenetrable, so I felt especially proud when I finally deciphered it, but it was hard to convey this fun in language anyone else could tolerate. Several times, I tingled with excitement at some new fact and wanted to share my joy. Over and over again, as I sat across a lunch table from friends, I watched their eyes glaze over while I tried to convey my new enthusiasm about thiamine. This is why I write books. I don't have to watch your eyes glaze over; I can't see you checking your watch. Trust me, thiamine is cool.

It started when I thought I read that thiamine helps our bodies produce hydrogen sulfide. Although I can't say that, there does seem to be a relationship but I only have a flicker of evidence.

One of hydrogen sulfide's evil talents is that it makes wine taste bad. Good wine can't have much hydrogen sulfide so wine makers work to reduce it. In a 2010 study, Bartra Casado and other scientists confirmed that the hydrogen sulfide in some inferior wines is produced by the yeast. They also proved that wine tastes much worse (way more hydrogen sulfide) when the yeast also makes too much thiamine. As they said in the Journal of Applied Microbiology, "This study provides a first hint which indicates that for some yeast strains, biosynthesis of thiamine (and perhaps of other sulfur-containing compounds) may be more relevant than the general nitrogen metabolism in explaining sulfide production by some yeast strains during vinification, defining new targets for genetic improvement of wine yeast strains."

That is, even if I imagined or dreamed the study that got me started down this path, it sounds like thiamine promotes the production of hydrogen sulfide in some way or another. At least in wine.

Cysteine contains sulfur and so does thiamine. In fact, the word "thiamine" comes from the Greek word for sulfur. When you heat either cysteine or thiamine, one by-product is hydrogen sulfide. Decades ago, scientists

169

decided that hydrogen sulfide contributes to the taste of cooked meat. Cysteine became an ingredient in artificial meat flavoring. They experimented with using thiamine as well, but I don't think it caught on.

You may remember they discovered thiamine as the cure for beriberi, the disease people get when they only eat white rice without the bran. In the mid 1800s they figured out that simply providing a more balanced diet prevented the disease. That was obviously a huge step, but it wasn't very precise. In the first decade of the 20th century, Casimir Funk formally proposed the idea that tiny amounts of some unidentified nutrients were necessary for health. In 1915, working with egg yolks, scientists figured out that some of these nutrients were soluble in fat (they called these "fat soluble A") and some were soluble in water ("water soluble B"). Scientists spent the next several years trying to identify them more specifically.

Their experiments seem a little ruthless today. They fed pigeons and chickens a diet of nothing but white rice. Within a couple of weeks the birds began showing signs of paralysis, which indicated the birds were deficient in the mysterious micro nutrient. Then they fed them various foods to see what cured them. The experiment had a narrow window. The correct nutrient, given on one day, would cure them and they'd recover. Given only a day or two later, the birds wouldn't die but would remain paralyzed. A day later was too late. They all died.

The nutrient turned out to be thiamine. We now know there are several varieties of thiamine; most are water soluble (like the kind found in rice bran) but some (like the kind found in garlic) are fat soluble.

Some parallel experiments during the same time tried to identify a mysterious growth factor in the diets of rats. When fed a diet containing every known dietary requirement (fat, protein, carbohydrates, etc.) in purified form, the rats failed to grow and reproduce. Adding tiny amounts of other foods solved the problems. These foods must have some micro nutrient as well. Once again, the magic ingredient was thiamine.

It was also the secret ingredient that reversed neuritis, an inflammation of many of the nerves in the body. The vitamin was finally isolated in 1933 and became a big deal. By that time, scientists had already isolated the factor in liver that promoted red blood cell production. Thiamine was the first one they were looking for, so it got to be called B-1, while the one they isolated from liver came to be known as B-12.

Since then we've learned that a thiamine deficiency can inhibit oxidative metabolism. That is, it can worsen the oxidative stress from superoxide and ROS. It can also worsen the results of inflammation. And it can make it more difficult for your body to deal with excitotoxins. Wait a minute. Scientists are studying three different aspects of toxins as contributors to ALS and Alzheimer's: oxidative stress, inflammation, and excitotoxins. And a thiamine deficiency could make any of those worse? Am I missing something here? Doesn't that seem like a fascinating clue?

Thiamine deficiency can cause nerve cell damage, that much we know. "Wernicke's encephalopathy" (or WE) is a severe thiamine deficiency common to alcoholics. Victims of WE suffer from confusion, lack of short term memory, incoordination, and paralysis of the eyes. Ultimately, it can lead to psychosis and death. The disease was first described in 1881, but we still don't exactly understand each step of the chemical processes involved.

We also don't exactly understand the biochemical processes at work in ALS, Alzheimer's, and Parkinson's Disease. But three suspects always seem to be near the scene of the crime: oxidative metabolism, inflammation, and excitoxicity. It would be useful if we could create a "model" to use to study these process, just like we breed laboratory animals to have conditions we want to study. A good model would be something we could duplicate easily but that we know how to cause and cure. Ideally, it would involve all three of these suspects.

In 2009, Alan S. Hazell and Roger F. Butterworth published a study in the Oxford University Press. Because thiamine deficiency (TD) "impairs oxidative metabolism," and it can lead to inflammation, and it can cause "excitotoxic-mediated cell death" they suggest it might make an excellent model .

As they state in their conclusion:

"Here we have highlighted oxidative stress, excitotoxicity, and inflammation in terms of our present understanding of their involvement in TD (Thiamin Deficiency). How these processes together help determine focal neuronal cell loss in TD and in cases of WE remains unresolved at the present time. However, what is now clear is that TD represents a useful model system for examining the interrelationships between these different mechanisms, as impaired oxidative metabolism is also a feature of neurodegenerative disease, and we know that several of these disease states also display elements of oxidative stress, excitotoxicity and inflammation"

They further suggest that thiamine deficiency, as a model, has "the potential to provide new insight pertaining to the pathophysiology of Alzheimer's disease, Parkinson's disease, amyotrophic lateral sclerosis, and other disease conditions in which these three processes are known to occur."

I had to read that a dozen times to decipher it, but I think I got it. If I understand correctly, they're saying that thiamine deficiency can act an awful lot like each of the diseases we're studying. That's just the deficiency alone. I wondered if it could act as a contributing factor if someone was also exposed to a toxin? I hadn't read anything to suggest it might be, but it interested me enough to continue learning about this mysterious substance.

In the organic steam engine of our bodies, we use oxygen to "burn" glucose. The process requires thiamine. The organs that use oxygen at the fastest rates are our brains, hearts, and livers. It's no wonder that those organs are most effected by beriberi.

Many foods contain thiamine and our bodies only need a tiny amount to prevent death by beriberi or paralysis from Thiamin Deficiency Syndrome (as it's known to tropical fish hobbyists). But, because the results of deficiency can be so devastating, it is often added to bread and cereal and other foods. It's included in every multi vitamin. A couple of scoops of brewers yeast contains eight times the recommended daily amount. If you're looking for a culprit in the American diet, thiamine has a perfect alibi. You'd have to make a conscious effort to avoid getting enough of it.

Or would you? Once you start looking at all the things that can contribute to a thiamine deficiency you start wondering how anyone in the world could possibly get enough of the stuff. For example:

Thiamine is involved in converting carbohydrates into energy. The more carbs you eat, the more thiamine you need. But how many of us adjust our intake of vitamins to match our intake of carbs? It's not added to soft drinks or candy bars or sugar or french fries. If you survive on soft drinks and candy and never take a vitamin or eat a vegetable, you might be a lot lower on B-1 than you realize.

It can be degraded or diminished by cooking (although acids like tomato sauce seem to protect it) and by exposure to chlorine or fluorescent lighting. It can decrease, over time, in stored food, including frozen food.

One common diabetes drug (metformin) may interfere with thiamine absorption. Yet many doctors don't suggest their diabetic patients take vitamin supplements to offset this.

Chronic kidney disease reduces the effectiveness of thiamine transporters.

Alcohol interferes with thiamine absorption. Alcoholics routinely have insufficient thiamine; so often, in fact, that thiamine deficiency in alcoholics has its own name, with entire books written about it. Thiamine deficiency may cause many of the dangerous side effects and discomfort of alcoholism.

Coffee and tea contain ingredients that can reduce your thiamine levels. The tannic acid in other foods can as well.

Your body doesn't store enough thiamine to last very long, so you need to replace it frequently. It is eliminated first in urine. When that route reaches capacity, your body will eliminate it in feces and in sweat. One way or the other, your body gets rid of any excess within hours. Because it's eliminated in urine, people who take diuretics are at risk of becoming low on thiamine. Low thiamine can lead to heart attacks, so doctors often prescribe extra thiamine for patients on diuretics.

But diuretics come in many forms and they don't all require a prescription. Coffee, beer, and alcohol can have the same effect. High blood pressure can act like a diuretic. Diabetics tend to urinate much more frequently than nondiabetics, as their bodies work to eliminate excess glucose from their blood. This may be why a diabetic's blood thiamine level is typically a fraction of a nondiabetic's.

It's possible to develop symptoms of thiamine deficiency by getting too much sulfur in your diet. At least, it's possible for cows to. In feedlots, cattle sometimes develop a disease called "Polioencephalomacia," which is mercifully abbreviated PEM. In this disease, part of their brain deteriorates and looks, upon necropsy, gray and limp. Symptoms can include blindness, lethargy, and seizure. Sometimes the poor cow just stands like a statue and stares. Guys who work in feedlots call these cows "brainers." Often, the disease is just a thiamine deficiency; one standard treatment is to give massive dose of thiamine. In many cases, if they catch the disease early enough, this cures the problem.

But other times the cows get these symptoms because they're getting too much sulfur in their food or water. It often happens when ranchers feed them too much of the by-products of ethanol production, the grain mash that's left after fermentation and distilling has extracted the most valuable components. Other times, the feedlot happens to have a water supply that contains too much sulfur. The bacteria in the cow's intestines produce too much hydrogen sulfide, their bodies can't detoxify it all. In these cases, massive doses of thiamine don't work. You have to change the diet or water supply.

So dietary sulfur might affect thiamine levels. That idea led me down a weird little detour.

Durian fruit, popular in the Philippines and other Asian countries, is most famous for its terrible smell. If you get past the smell and actually eat the fruit, it's supposed to be fabulous. Durian has a very high sulphur content. John Maninang and Hiroshi Gemma from the University of Tsukuba in Japan got interested in the legend that eating durian fruit and drinking alcohol will make your stomach explode. They did some experiments. It turns out that durian juice interferes with the body's ability to process the toxins we manufacture as we break down alcohol. Not just a tiny bit, but by up to 70 percent. Their 2009 paper suggests that eating durian fruit and getting drunk might kill you, just like folklore suggests.

Although I don't think durian grows on Guam, it's an example of eating something with excess sulphur making it harder for your body to process a relatively harmless toxin. By itself, durian is considered a nearly magical health food, full of vitamins, calories, fat, and nutrients. In combination with alcohol, it's dangerous. Interesting. Let's put that clue on a three by five card and return to things that might reduce thiamine in a typical American.

Sulfites (another chemical group that contains sulfur) are used as preservatives in food. They destroy thiamine. If you're eating food that's high in sulfites, you'll need to get more thiamine into your diet. Do you keep track of the sulfites in your diet? Me either.

Thiamine requires magnesium to do its work. If you're low on magnesium, your thiamine will be rendered ineffective. In a cruel twist, excess alcohol not only makes it harder for your body to absorb thiamine, it can also cause a magnesium deficiency. Interestingly, on Guam scientists investigated low calcium and magnesium levels in the drinking water. They thought that might lead to conversely large aluminum uptake by the natives. Let's move that three by five card to the top of the pack.

Let's summarize the things that can lead to a thiamine deficiency: a diet that's high in any of the following: carbs, sulphur, alcohol, coffee, tea, or sulfites (from processed foods, for example). Food that's overcooked. Taking certain medicines, especially for diabetes. Taking diuretics, or having a condition that causes excessive urination. Being low on magnesium. Having chronic kidney problems. Being older.

Are you still completely confident you get enough vitamin B-1?

Wait, there's more.

Thiaminase

Some things actively counteract or destroy vitamins. They used to call these factors "antivitamins." They sometimes still call them antagonists. These chemicals work in two different ways. One kind binds to the same sites as the vitamin, but doesn't do much. There are only so many chairs in the choir and if they get filled up with nonsingers, you won't have much music. Chemicals that replace a vitamin effectively reduce what the real stuff can accomplish.

The other kind of "antivitamin" actually destroys the vitamin. One of these is called "thiaminase." It's an enzyme that shatters the thiamine molecule into useless pieces.

Here's a brief vocabulary tip before we continue: the suffix "-ase" attached to a chemical describes an enzyme that breaks down the chemical. A "proteinase" (or "protease") breaks apart a protein. "Thiaminase" is an enzyme that breaks apart thiamine.

Thiamine is found in many foods and in several forms. Thiaminase also comes in several forms and is found in various foods. Beets and blackberries have small amounts of it, for example. As long as you're getting enough thiamine in the first place, a little thiaminase won't hurt you. But some foods have so much they're dangerous. Luckily, cooking destroys most kinds of thiaminase.

There are at least two varieties of thiaminase. Both can be manufactured by bacteria that sometimes live within the human intestine. According to L.L. MacDowell, in his book *Vitamins in Human and Animal Nutrition* (2000):

"Some Japanese studies (Zintzen, 1974) have indicated that structure-altering antagonists of thiamin…produced by bacteria and fungi are main causes of beriberi, since they are involved in up to 70 percent of all cases. Bacillary and clostridial species that produce these antagonists have been isolated from intestinal flora of Japanese people tested. They may be consumed with contaminated food, such as moldy rice."

Some foods contain dangerous levels of thiaminase.

Remember the bracken fern? It blinds, paralyzes, and kills livestock and humans — but pigs seem to love it. Its "poison" is thiaminase, which destroys the thiamine in their bodies. Pigs tolerate it because, unlike humans and horses, they store a whole bunch of thiamine in their tissues. Cattle tolerate eating small amounts because they have such a magnificent digestive system with multiple stomachs and gloriously long intestines. The bacteria living within them produce lots of thiamine. Even at that, bracken can sicken a cow or pig that eats too much.

This thiaminase stuff doesn't fool around. The first British expedition that tried to cross Australia from south to north learned this the hard way in 1860. When they ran out of food, they decided to make flour out of the native "nardoo" fern just like the local population did. But they didn't follow the detoxification instructions closely enough and three men died from thiamin deficiency.

You and I may not have heard of thiaminase until this week, but it has been the subject of much study and conversation for decades in three different communities: people who manage fisheries, tropical fish hobbyists who raise certain kinds of fish, and people who keep reptiles as pets.

Many kinds of fish, both freshwater and saltwater, contain thiaminase. Because most kinds of thiaminase are destroyed by cooking, it's not a huge deal for humans. But if you've got a big aquarium full of predatory fish, like piranha or gulpers, thiaminase may be your biggest enemy. Many predatory fish only eat fresh, frozen or live food. They won't eat flakes or pellets. If your feeder fish contain much thiaminase, your pets will become thiamine deficient over a period of days or weeks and finally die. Freezing doesn't destroy thiaminase. Unfortunately, it's an enzyme that's found in nearly half the varieties of fish, shrimp, and bivalves hobbyists can easily obtain. Goldfish, carp, and mussels seem to have more than other varieties. In the wild, this isn't usually an issue because food supplies also contain lots of thiamine. In captivity, with a more restricted diet, you run the risk of providing a daily diet rich in thiaminase but deficient in thiamine.

Other kinds of exotic pets face the same danger. I wish I'd learned this years ago.

When I was a kid, pet stores occasionally sold "baby alligators." Technically, they were often "caimans," a close relative. Mine was a constant primeval delight. I felt like I had a miniature pet dinosaur, a living bundle of crude instincts and bad temper. He seemed so primitive and ancient it didn't feel right to name him. He came from a time before names existed.

In the summer, I'd catch grasshoppers and earthworms for him to attack and devour. I never tired of watching his single-minded hunger. When winter came, I could sometimes get him to attack a bit of raw liver at the end of a broom straw. But mostly I fed him goldfish the pet store sold cheap for just this purpose. I figured that was probably the closest I could come to providing him the diet he'd have had in a Florida swamp. I'd never heard of thiaminase and I'm sure the folks at my local pet store hadn't either. He clearly relished the goldfish.

If you've watched an alligator on a TV special attack an antelope, clamp down, and shake its prey, then you know exactly what it looked like. But gradually, over the course of two or three weeks, he became lethargic and finally died. Fifty years later, I may finally understand why. The goldfish contained thiaminase and I wasn't providing enough dietary variety to compensate for it. My pet probably died of thiamine deficiency.

In the wild, fish and alligators usually eat a varied diet. They get enough thiamine to overcome the thiaminase. In some places, however, limited food choices force them into a diet without that balance. Salmon in the Baltic Sea wind up subsisting almost exclusively on little fish called alewives that have too much thiaminase. The salmon get sick. Some die, others manage to survive and lay eggs but then the eggs don't hatch, or the fry don't live long. This problem threatens the entire fish population in the area.

Thiaminase is a major problem in the Great Lakes, but we're still investigating its source. The invasive zebra mussel and the quagga mussel, accidentally introduced a few years ago and first noticed in 1991, can contain concentrations 100 times greater than fish in the same environment do. They're only about the size of your thumbnail, but one adult can produce as many as a million eggs in a year. Any pest with that reproductive enthusiasm is hard to eliminate. Most fish don't find them palatable, but a few have developed a taste for them. Could they be the source of thiaminase in the food chain?

We don't know.

Glutathione

Perhaps the most common antioxidant in our bodies is glutathione. It's found in virtually every cell of our bodies, but older people don't have as much. Not only does it act as an antioxidant itself, it helps recycle vitamin C and E so they can be more effective. Many people believe that a glutathione deficiency plays a huge role in neurological diseases. If you decide to research it yourself, you'll find thousands (perhaps millions) of references to it online.

We don't need to eat glutathione, because our bodies manufacture it. In fact, if we try to increase our supply by eating the stuff, we just digest it into component parts without increasing our levels. Still, many companies will sell it to you. Short of getting glutathione injections (which is being tried experimentally with mixed results) the best strategy for increasing the amount in your blood and liver is to give your body the raw materials to manufacture it.

Precursors of glutathione include N-acetylcysteine (often called NAC) and S-adenosylmethionine (SAMe). Interestingly, whey protein is also believed to increase levels of glutathione. These are all readily available.

To manufacture glutathione, we also need at least two other ingredients that act as catalysts or co-factors. One of these is calcitirol, which our bodies make from vitamin D. The other co-factor is thiamine. The first sentence of a study published way back in 1960 by scientists Hsu and Chow at John Hopkins University was: "Thiamine deficiency results in decrease in concentration of glutathione in erythrocytes and heart, but an increase in level of glutathione in liver tissue."

If you don't have enough thiamine, your heart and blood won't have enough of your main antioxidant.

People now sell glutathione accelerators, pills that claim to increase your body's production of the antioxidant. Some people rave about the results they obtain. The supplements probably contain the raw materials and cofactors our bodies need to manufacture it. I haven't studied them or tried them and the manufacturers are quite secretive about their ingredients. Some are probably better than others, but the science behind the concept seems promising.

Personally, I take a daily vitamin D and B1.

Back In My Basement For a Moment

I've seen Suzanne several times since the last time I mentioned her to you. One time, I brought her some vitamins and fixed her vacuum cleaner. Another time, we drank altogether too much wine with our friend Cathy. We all laughed a lot and it was great fun. My wife and I attended a party Suzanne's friends held to raise money for the ALS Association.

Unfortunately, her disease has progressed. Now she can't hold a wine glass but, trooper that she is, she has adapted to drinking through a straw in a covered plastic cup. The logistics of her changing circumstances occupy most of Suzanne's time; she'd like someone to find a cure, but the research that leads to it does not fascinate her the way it does me. She did point me toward one intriguing new area of research. Recent studies suggest that marijuana may actually have therapeutic value for ALS, beyond its calming effect. I have no idea why it might work, but the evidence is fascinating and I bet it would be easy to get volunteers for that study. In Colorado, medical marijuana is legal and Suzanne, always willing to be a pioneer if she can help humanity, tells me that it's about the most useful drug she's been prescribed.

I raised 46 baby catfish to near maturity and traded them to a pet store for credit against future purchases. I'm now raising another two dozen.

My three surviving jackfruit trees are about two feet tall and thriving. They spent the summer outside in partial shade, then I brought them in before winter. It turns out, the fluorescent light above my desk provides them a fine amount of light. I bought some mycorrhiza spore at my local nursery and added it to their pots. The leaves are now twice as large as they were before and gleam with a healthy shine. The plants seem very happy.

The bracken fern in my terrarium became the most lush vegetation I've ever maintained indoors. Just as I'd imagined, it looked like a scene from a prehistoric jungle. Delicate little mushrooms sprouted from the soil every few days. But then, mysteriously, the bracken began to droop. The soil seemed a bit dry, so I watered it. But it just kept getting worse. Could the micorhyzzal fungi I added to help some tree seeds I'd planted have done the damage? Did some other pathogen attack it? I don't know. I did learn that bracken produces herbicides to kill competing plants and that some varieties of bracken act like deciduous trees and lose their leaves in the fall. Maybe it's just resting for the winter. Maybe some spores will sprout. I won't do anything except keep the soil moist and wait.

It may surprise you to learn that I sometimes read books just for fun, both fiction and nonfiction. Most of these have nothing to do with any writing project I'm involved with; I just like to read. Sometimes they lead to a new project or hobby. One of my sons recently gave me a book on making cheese, for example. Cheese making looks like a reasonable thing to attempt, plus I learned some new bits of everyday science. I had not realized that both bacteria and fungus are involved in the process of making many varieties of cheese. I thought it was just bacteria.

I also just read *The White Zone,* a 1994 book by Richard Preston about the first terrifying outbreaks of Ebola and Marburg virus. A friend recently suggested I'd enjoy it, so I was familiar with the title. This Christmas, one of my sons, by sheer coincidence, gave it to me as a gift. I did enjoy it, although it's pretty scary and has been banned from some high schools for being too graphic. In another weird coincidence, that book circled around to some of the things I had been thinking about for this book.

Ebola and Marburg are nasty viruses. Within a week or so of exposure, a person's body pretty much disintegrates. The victim bleeds from every orifice and soon dies. The virus passes from monkeys to humans (and from human to human) through exposure to infected blood or flesh. In at least some cases, it can spread through the air. In a bizarre twist, one variety (now known as the Reston strain) is deadly to monkeys but doesn't really seem to bother humans much. Maybe the deadly form mutated in a happy way for humans, maybe the benign form is just waiting to mutate into the deadly form. I don't want to think about it too hard.

The diseases kill monkeys very quickly, which is a bad strategy for a virus. It's dumb to exterminate your food supply. Smarter to prey on animals that survive long enough for you to infect many other hosts. Therefore, the virus might normally live within some animal that tolerates it better and occasionally "jump" to humans or monkeys.

In the final chapter, the author decides to visit a certain African cave that seemed to be a common factor in the first reported cases of these diseases. The reader gets nervous right along with the author as he dons protective gear to explore the cave wondering what animal living there might be the one that harbors the virus before it leaps into a human. He notices the birds outside, the elephant markings, the spiders, the insects. Any one of them might be the carrier. There are dozens of possibilities. He's very careful and breathes through a filter, but you can feel his heart beating faster.

My own heart started beating faster when he casually reports seeing fruit bats in the cave. Look up! I wanted to scream at him. It's the fruit bats! Get out of there! But, of course, that was just paranoia. I've been reading so much about fruit bats and cyanobacteria that I've developed a little phobia about them that isn't at all scientific. When the book ended, no one knew what animal might be the missing carrier of the virus, or even if there was one.

Still, I wanted to be sure, so I did some research.

In the two decades since that book came out, scientists have studied literally thousands of animals searching for traces of these deadly viruses. They have discovered within the last few years that the Ebola and Marburg viruses do, in fact, live in African fruit bats. One fruit bat that tested positive for the virus lived for over a year without ever showing any symptoms.

Then, in 2008, Ebola virus was reported in pigs in the Philippines. In 2009, a pig farmer caught a strain of Ebola from one of those pigs. Luckily, it was the Reston strain which is deadly to monkeys but relatively harmless to humans. There has been speculation that the pigs got the disease from fruit bats, perhaps in bat saliva left on a partially eaten fruit the pig found.

I report this only because it felt very weird to stumble across fruit bats in this entirely different context. It certainly demonstrates that fruit bats can survive some things that are deadly to other species. It would be interesting to understand exactly why different species respond to toxins and virus in such dramatically different ways.

Maybe there is some tool in another species' toolbox that humans could adapt to deal with similar threats.

Thiamine Trivia

I read many odd things about thiamine. Most have nothing to do with our quest, but I'm a guy who can't resist sharing cool facts. Perhaps I don't need to tell you that at this point. Some very impressive scientists believe thiamine deficiency (which they abbreviate TD) is implicated in various neurological diseases, especially the ones that are more common among older people. Who knows what weird little tidbit might be an important clue?

I doubt this one is important: Many people believe that thiamine is a secret weapon against mosquitoes. Some seemingly sane people swear by this. They claim to get near perfect protection against mosquito bites just by taking a thiamine pill a half hour before they go outside. They claim that mosquitoes hate the smell of the trace amounts their body eliminates through sweat. This makes some sense; that "vitamin" smell you notice when you open a fresh bottle of multivitamins comes from thiamine.

This idea is so prevalent that they've actually conducted some tests. In each test, the results have been inconclusive. I would suggest that maybe both sides are correct. Maybe the smell of thiamine does repel mosquitoes, but your body isn't going to eliminate it via sweat until you've stored as much as your body thinks it needs. If you're a little deficient to begin with, your body will absorb what it can, then eliminate the excess via urine. I think it would be interesting to try the experiment, but make sure the participants weren't at all deficient to begin with. Or maybe just dab a bit onto their skin. If I had some mosquito larva handy, I'd put some thiamine in with them right now just to see if they reacted. As luck would have it, I'm all out of mosquito larvae, so I guess you'll have to try it instead. If it did work, the really interesting question would be, "why?"

Plants, fungi, and bacteria produce thiamine. Animals can't. We believe our common ancestor lost that capability millions of years ago through a genetic fluke. The theory is that losing the ability to manufacture our own B-1 didn't hurt animals because it's so easy to get thiamine from food. (A similar theory, made famous by Linus Pauling, explains why a few animals, like humans, can't make their own vitamin C. In that theory, our distant ancestor lost the ability to make Vitamin C in a genetic mutation. Our loss of self-produced vitamin C happened only 25 million years ago, rather than 500 million years ago for thiamine.)

After I described what I'd learned about B vitamins, a couple of friends started taking them, especially thiamine. After a week or so, one complained that she was sleeping so long (18 hours one night) she gave it up and her sleep patterns returned to "normal" which, for her, is maybe six hours a night. So I looked up thiamine and sleep. In a paper from 1982, scientists (Crespi and Jouvet) reported that thiamine deficiency made mice sleep less. When the deficiency was corrected, they sometimes over shot the mark and slept too long. Interesting.

The same search led me to a number of people who believe that anorexia is a form of thiamine deficiency, as well as some that suggest thiamine deficiency causes nightmares. This led to the Ghost Death of the South Pacific. Since 1917 there have been dozens of deaths among young men in the Philippines, Japan, Vietnam and Guam. They moan in their sleep, as if having a terrible dream. Those that recover describe the sensation of a ghost sitting on the chest and paralyzing them. The less fortunate die when their heart stops beating. It's called Sudden Unexplained Death During Sleep and abbreviated SUDS. In the Philippines, it's called "bangungut." It often occurs after a young man eats a large meal of white rice and fish. Several scientists have pointed at a thiamine deficiency as the ultimate cause. Something similar has been reported for generations all over the world in a less specific way. It may be what inspired the term "nightmare."

In 2011, a study showed that "Insulin Like Growth Factor II" (IGF-II) improved memory. Thiamine is involved in our body producing this chemical. Another study showed that inhaled insulin improved Alzheimer's. Diabetes is an insulin-related disease and diabetics have dramatically reduced thiamine.

Like BMAA, thiamine has been considered before. In small studies, Alzheimer's patients treated with thiamine got little benefit. On the other hand, some feel that thiamine treatments need to be massive and prolonged to overcome damage caused by a deficiency. It would have been great if the studies were more positive. But my own line of thinking was not that B-1 might be a cure, but rather that a deficiency might be a contributing factor.

Working by myself, I didn't find much confirmation of that theory online. But once I started sending portions of this book out to experts, I learned that, once again, I came late to the conversation. After Derrick

Lonsdale, M.D. (who is on the Scientific Research Advisory Committee of the American College for Advancement in Medicine) graciously read through the thiamine sections, we spoke on the phone. He corrected a few of my sentences. I asked if it would be reasonable for me to say that it's *possible* that thiamine deficiency could be a contributing factor in these neurological diseases. He indicated that, not only would that be true, but he thought it was an understatement. He referred me to several more books and scientists. It's obvious I've just scratched the surface.

Back in 2001, Gary E. Gibson and Hui Zhang of the Weil Medical College published a 12 page paper on the role of thiamine and oxidative stress. The first sentence of the abstract is, "Thiamine-dependent processes are diminished in brains of patients with severe neurological diseases." It goes on to list various puzzles that remain to be solved about the processes. Then it says, "The data indicate that the interactions of thiamine-dependent processes with oxidative stress are critical in neurodegenerative processes."

The tone implies (to me, at least) that they don't consider thiamine deficiency to be some wild, speculative goofy idea. It sounds like many scientists accept that a lack of thiamine either plays a role directly, or is at least an intriguing clue to the inner workings of these diseases. But anyone can publish on the Internet. Was this "Weil Medical College" a reputable source? I went to its website.

Here in Colorado, I'd never heard of the college, but I had heard of the university it's part of: Cornell University. OK, Cornell is pretty reputable. According to the Weil Medical College website, "Gary Gibson and his research group are trying to discover the underlying cause of age-related neurodegenerative diseases and to develop effective therapies." It goes on to list four specific aspects of brain chemistry they intend to focus on. One of the four is thiamine deficiency. In November, 2011 the college website announced that it had received a 100 million dollar gift to further its work. In five years, it's received over a billion dollars. Somewhat sheepishly, I decided that these are not small time kooks with strange ideas. While they're doing their research, I will quietly take my B vitamins.

The chemistry and math in these studies is beyond my brain, so my mind wandered to areas where I feel more comfortable.

We know that thiaminase can trigger a sudden, deadly deficiency in an otherwise healthy individual. It's a big problem in the commercial fishing industry and I'm interested in fish. I wondered how thiaminase became such a problem for the salmon industry? It made sense to start with their diet.

Thiaminase in Fish

I googled the alewife, the fish in the Baltic that contain way too much thiaminase. I wondered if they could be getting it in their diet and concentrating it the way Paul Cox suggests fruit bats concentrate the toxin BMAA. I wanted to know what alewives eat.

One study examined a whole bunch of alewife carcasses; most of the stomach contents were copepods, nutritious little shrimp-like critters. If you have a salt water tank, you can actually buy a little bottle with over a thousand microscopic copepods to start a colony of your own. Alewives may eat other things as well, but it looked like these were their favorites. Copepods are considered zooplankton; that is, animals that live in and make up part of plankton. So what do copepods eat?

Remarkably, these little fellows have been studied extensively; scientist know a lot about copepods. I just didn't happen to, so I read several articles about them. Their diet was easy: copepods feed on algae.

Of course, I wondered if copepods ever ate blue-green algae? Would they be smart enough to avoid cyanobacteria, or had the two life forms evolved beside each other for so many eons they had developed coping skills?

It turns out that copepods do, indeed, eat cyanobacteria. But, for critters the size of a speck of dust, they're pretty smart about it. Somehow they can tell which ones are full of toxin and avoid them. Perhaps they can smell the difference. (OK, I don't know if copepods have a sense of smell, but then you don't either.) Scientists have done specific experiments to test this and they're confident. Copepods avoid toxic cyanobacteria and eat the nontoxic ones.

Unless they're very hungry. If it's been long enough since its last meal, a copepod forgets to be selective and will eat whatever cyanobacteria you put on its plate, toxic or not. If the copepod happens to find itself in the middle of a blue-green algae bloom, and that's all there is to eat, he's gonna eat it.

If we were investigating high levels of BMAA in alewife fish, we'd be pretty excited about that. But we were looking for a source of thiaminase, not BMAA. And cyanobacteria don't manufacture thiaminase— or do they? That would be too weird, wouldn't it? Still, I couldn't resist looking it up, just to be thorough.

In article after article I found the same thing: "cyanobacteria produce extensive amounts of thiaminase."

I stared into space. That was just too weird a coincidence. I wondered if the cyanobacteria in cycad roots produced thiaminase as well as BMAA? I wondered if anyone had tested the effect of both chemicals on a person? If they did, I never found it. Then I started thinking about the problem of the salmon industry and how thiaminase might work its way up a food chain.

Those zebra mussels and quagga mussels in the Great Lakes are filter feeders — they take in whatever algae, plants or zooplankton happens to float past them and expel what they don't want in a mucus covered blob called "pseudofeces" before it ever gets to their digestive system. Their bodies become concentrated with various toxins from their food and so do their pseudofeces. Maybe they collect thiaminase from cyanobacteria in their bodies the same way, just like fruit bats concentrate BMAA. I don't know if that's what happens, it's just an idea. Still, a salmon wouldn't need to eat too many mussels or pseudofeces to get sick. It could even eat a few accidentally.

Humans can become thiamine deficient in so many ways we'd never notice a common pattern among victims. If thiamine deficiency actually was a contributing factor, along with some toxin, it would be a very complex puzzle to solve. On the off chance that there's something to that, I take my B vitamins and don't worry about it too much.

But I look at sushi and raw oysters with newly suspicious eyes these days.

I confess, the idea that cyanobacteria produce thiaminase startled me. I'd almost say it frightened me. At the least, it felt like a sign that I need to finish this project and move onto something else. I had been concentrating on tracking friendly vitamins and antivitamins, eyes down and focused on the trail ahead, when suddenly a huge gob of blue-green algae leaped from behind a bush and yelled "Surprise!" I thought I was done with them.

Now I had to think about them all over again.

I've Decided I Don't Like Cyanobacteria

Cyanobacteria do not seem like our friends, even when you call them by the cuter name of "blue-green algae." We don't really know all the toxins and antivitamins they create; they certainly seem to be operating below our collective radar. It would be smart to stop coddling them. That is, we need to eliminate the excess fertilizer we provide them in the runoff from our lawns, fields, and feedlots. That green muck in the pond at your local park is not lovely evidence of nature at work. It could be this ancient assassin plotting against your children and pets. It might be robbing your grandmother of her memories and your cousin of his ability to walk. It hasn't been proven, but is it smart to wait until all of us live in hospitals and nursing homes before we pay attention to the possibility? Many smart businessmen have read the same stuff I've read and are already out there, ahead of the rest of us, hoping to make a huge profit. A couple of examples:

A recent study suggested that you could not only kill blue-green algae but also detoxify its toxins by using ultrasound at a specific frequency. That sounds goofy to me, but some entrepreneurs got right on it. A Dutch company has begun manufacturing ultrasonic water treatment equipment. Sales have been so brisk they opened an office in Atlanta. They use them to clean swimming pools and lakes. My son Scott suggested detoxifying people by letting them soak in a jacuzzi with ultrasound. And maybe some mud. Don't know how that would work, but it sounds good to me.

Another company sells solar powered water pumps that float on a pond or lake and circulate the water, increasing its oxygen content and making it less comfortable for cyanobacteria. Their business seems to be booming. There may be a market for something similar for livestock troughs. If I had some cows or horses, I would not want them drinking water full of blue-green algae.

Some people put a little bundle of barley straw in their ponds, because it releases hydrogen peroxide as it decomposes, which kills cyanobacteria.

Eliminating excess nutrients from warm, still water is the big step. But there may be many smaller strategies for reminding cyanobacteria that we have our own claim on their ancient territory. Some of those will probably prove to be very lucrative.

Meanwhile, I'll keep watching over my shoulder for gooey blue-green protoplasm stalking me in the night.

Thiamine and Guamanians

Last night I woke up at about two wondering about something. This project started with cyanobacteria and the Chamorros of Guam. It's come back around to thiamine and cyanobacteria. But what about Guam? Did the Chamorros have any issues with thiamine?

Turns out that, until the 1940s, the biggest health issue for people living in Indonesia, Japan, the Philippines (and presumably Guam) was thiamine deficiency. They ate white rice, polished clean of the bran that contains thiamine. In 1946, R.R. Williams, the guy who figured out the formula for thiamine, came up with the idea to fortify rice with vitamins, including thiamine. He started with a large pilot program in the Philippine province of Bataan. Within two years, the area had dramatically fewer cases of beriberi and many fewer deaths. Because of that success, in 1952, the Philippines passed a law that required all rice mills to fortify their product with vitamins and minerals, including thiamine. Unfortunately, the mill owners hated the law and many refused to comply. Still, cases of beriberi declined. The problem is much smaller today, but not gone altogether. According to a paper published in the Journal of the American Dietetic Association in by Leon Guerrero RT et al, as of October, 2009, white rice sold in Hawaii, Guam, and Saipan often lacks nutrient enrichment. According to their study, "Rice that was labeled as enriched in Hawaii and Guam seldom met the minimum enrichment standards for the United States." They warned nutritionists studying the region not to assume people were eating enriched rice just because that's how the rice bag was labeled.

I find it interesting that the cases of lytico-bodig declined at exactly the same time that rice mills started adding thiamine to their products.

Prior to 1952, it seems likely that Guamanians suffered from thiamine deficiency just like everyone else in the Pacific region. The Chamorros probably had it even worse because some raw fish contain thiaminase and they loved raw fish. The first Europeans to visit Guam in the 16th century took note of this, including the missionary Fray Antonio del los Angeles. He reported that the natives ate raw shellfish, including mussles, and other raw, whole, ungutted fish. Interestingly, the internal organs modern fisherman discard are exactly the ones that develop high concentrations of thiaminase. Chamorro traditions continued unchanged well into the 20th century. If you're hiding in the jungle from soldiers and always hungry, and you've been eating raw fish your whole life, you probably wouldn't think much about catching whatever fish you could and eating it on the spot.

You would think scientists would have looked at thiamine deficiency among victims of lytico-bodig long ago. Actually, they did. In 1992, the Journal of Neurological Science published a study by Laforenze et al titled, "Thiamin mono- and pyrophosphatase activities from brain homogenate of Guamanian amyotrophic lateral sclerosis and parkinsonism-dementia patients." In case you missed the article for some reason, it compared the amounts of thiamine in the brains of Guamanians who died from ALS and those who died of other causes. Of the two forms of thiamine they tested for, one form showed no difference between the two groups. The other form was dramatically reduced in the brains of those who died from ALS. Their brains were deficient in thiamine.

ALS is not beriberi, but the two diseases do have some similarities. I wondered if maybe a combination of things might cause the disease, which would help explain why its cure has been so elusive. For example, could BMAA and other toxins be one part of the problem but a deficiency in thiamine weaken resistance to the toxin?

Maybe that's not so far fetched. In his book *Vitamins in Human and Animal Nutrition*, (2000) L.R. McDowell refers to a study that indirectly supports the possibility. He says, "Exposure of thiamin-deficient mice to ethanol resulted in brain damage that was more severe than either treatment alone (Phillips, 1987)."

As we age, our bodies require more thiamine for the same amount of food eaten. It becomes harder for us to absorb it through our intestinal walls. There are so many ways we can become deficient, including just having the wrong set of bacteria in our intestines. And remember, until the last five years or so, the test for thiamine in our systems was flawed. None of the science before 2008 was conducted by people who had access to the improved test.

It would be reasonable to make sure that anyone at risk for these diseases is not also deficient in thiamine.

Goats and Pigs, Iodine and Thiamine: a Hypothesis

It's easy to decide that iodine deficiency plays some role in neurological diseases like ALS and Alzheimers. For over fifty years we've known that you're much likelier to get one of these diseases if you spent your infancy in a region that's deficient in iodine. Another example: exposure to the fungicide maneb dramatically increases the chances of getting a neurological disease from a toxin; maneb works by disrupting the use of iodine in animals. Our instincts shout that iodine must be involved; but we can't say it out loud. Although iodine deficiency remains the number one cause of mental retardation in the world, we can't prove it has a role in Alzheimer's or ALS. No one has found a smoking gun.

A deficiency in thiamine (commonly known as Vitamin B1) is also a tempting suspect. For nearly a hundred years we've known that a severe B1 deficiency causes confusion, memory loss, paralysis, and death. The disease (called beri beri) includes all the symptoms of the worst neurological diseases. And the effects can be startlingly fast, first paralyzing and then killing the victim within days. Many scientists are scrambling to prove that thiamine deficiency is involved in Alzheimer's and ALS; some are confident they're close to a solution. But humans can store B1, we get it in our diet, much of our food is fortified with it just to make sure. How in the world would a modern American become so deficient he would develop neurological symptoms? It just doesn't seem likely. Something's missing.

I got to wondering about one aspect of thiamine that seems a little weird: the storage of the vitamin in our bodies. Animals, including humans, can't manufacture thiamine, so they have to get it from their diet, or from the bacteria in their intestines. A human can store enough to last anywhere from one week to a month. After that, unless it's replenished, they start showing symptoms. A cow stores more; beyond that, bacteria in its complicated digestive system churn the stuff out. Unlike other animals, pigs store a lot of thiamine in their muscle tissue. In one experiment, young pigs thrived for two months without any thiamine in their diet. At that point, the pigs showed no side effects so they stopped the study; we don't really know how long a pig can survive. Pigs seem to be the champions of thiamine storage.

But goats can't store thiamine. Luckily, the first of their four stomachs teems with bacteria that produce it. A goat might have 200 times as much thiamine in its blood as its human caretaker. As long as a goat keeps eating, the bacteria provide it with all the B1 it needs. But if a goat stops eating, or eats something that destroys thiamine (like moldy straw) or eats too much grain, it comes down with "goat polio." Like the human B1 deficiency, beri beri, the symptoms include confusion, "star gazing," and ultimately paralysis and death. Without thiamine, a goat can die within a day or two. That's right: within only a day or two.

Thinking about this in the middle of the night, I got an idea: if something mucked up a human's ability to store thiamine, that might cause a sudden case of beri beri including symptoms ranging from depression, confusion and memory loss to paralysis and death. I tried to find an article that explained the mechanism of thiamine storage, but couldn't find one. What I did find was just as interesting: when people suffer from an overactive thyroid, so they have too much thyroxine in their system, they can't store thiamine.

That's right. In those fairly rare situations where a thyroid gland pumps out too much product, one of the results is that humans can't store Vitamin B1.

At first this just puzzled me. People point to an iodine deficiency when they talk about a connection to neurological diseases. An iodine deficiency causes the thyroid to under produce, not over produce.

But there's one exception to that. When people have been deficient in iodine for a long time, their thyroid works extra hard trying to compensate. It can work so hard it swells until it's a visible lump in a person's neck called a goiter. This is a huge problem in third world countries. The cure is simple: give the patient iodine. But that swollen thyroid gland has been training for a long time to work very hard. If you provide too much iodine right away, the thyroid acts like a revving race car whose brake is suddenly released. It will flood the body with thyroxin to the point the patient can suffer a heart attack.

Here's the hypothesis. I don't know that this is true, and maybe other people have thought of it. It's just an idea:

Could something similar happen in people who have had a milder iodine deficiency for a long time? Maybe not diagnosed; the patient probably has no visible goiter. Just a thyroid that's been working hard against the emergency brake for too long. The patient probably feels tired all the time and tends to gain weight. Then, he changes his diet. Maybe he starts eating seafood, or iodized salt. Maybe he moves to a region that isn't deficient in iodine. For whatever reason, he now has enough iodine and his thyroid gland explodes with joyful production. Suddenly, his body has more thyroxine than it knows what to do with. He probably feels more energy than he has in years, so he doesn't complain. What he doesn't know, in this hypothesis, is that now his body can't store thiamine. If he gets it in his daily diet, fine. But perhaps his new diet doesn't include enough. Or maybe by evil coincidence, he eats something that destroys thiamine before he can absorb it. This combination of events would be rare, but possible. Within a very short time, maybe less than a week, he starts showing symptoms of a

B1 deficiency. Routine blood tests wouldn't catch this. In fact, before 2008 blood tests failed to accurately measure thiamine levels in the blood.

The original experiments that led to the discovery of B1 noticed that the damage from thiamine deficiency was reversible when caught in time. Chickens showing signs of paralysis made a full recovery when they got the vitamin on one day. A day or two later they would survive, but remain paralyzed. After a certain time, the disease was fatal. So, if this little theory is correct, patients who ultimately got some B1 might recover or not, depending on the timing. That would make it much more difficult to notice the connection. I can see why people might overlook it. By the time anyone checks for thiamine, the levels could be normal again, with the damage already done.

Consider the natives of Guam (known as Guamanians or Chamorros) in the 1940s. They came down with the symptoms of ALS at such an alarming rate that "lytico bodig" was the number one cause of death. But why? Scientists have focussed on the toxins in the cycad nuts they ate, which can cause these symptoms, but only in high concentrations. Maybe there was an additional contributing factor. Thiamine deficiency was considered the largest health problem in the South Pacific (presumably including Guam) until mid 20th century. But the lytico-bodig disease seemed to lie dormant for a long time and then strike suddenly, sometimes years after a patient had left Guam. That's not typical of a vitamin deficiency.

When Japan occupied Guam in 1942, it commandeered the natives' food and forced many Guamanians to work growing food for the japanese military. The plan failed; the island couldn't grow enough food to support all the Japanese soldiers, let alone feed the natives what was left. Chamorros were chronically hungry; many hid in the jungles for years and had to subsist on whatever they could find. Was their diet deficient in iodine? We can't know, because the very first study of iodine on the island wasn't begun until 2012 after scientists discovered that the islands of Fiji and Vanautu were deficient in iodine. If the results have been published, I haven't found them.

If the plants and animals of Guam had the same low levels of iodine as Fiji, the Guamians who spent years in the jungle may have developed thyroid glands that worked very hard to compensate. When they came out of the jungle and caught some fish, they may have created a perfect storm. The seafood contained a lot of iodine, so their thyroid glands shot into overdrive. The excess thyroxine may have prevented them from storing whatever thiamine they ate. Plus, they ate the fish raw, and many fish contain thiaminase. Unless destroyed by the heat of cooking, thiaminase breaks apart thiamine molecules in the body. The final fatal puzzle piece may have been the toxin BMAA which they got from eating cycad nuts and fruit bats. By itself, the concentration of toxins may not have induced paralysis. But in combination with the sudden thiamine deficiency, maybe it was.

In 1952, the Philippines (the largest rice exporter in the region) passed a law mandating that all its rice must be fortified with thiamine. No Guamanian born after 1952 has been diagnosed with "lytico-bodig."

Perhaps that's not just a coincidence.

Three Schools of Thought

Some scientists consider Alzheimer's, ALS, and Parkinson's Disease to be primarily genetic diseases. They have excellent reason to believe this. Statistical evidence supports the idea; they can sometimes predict who will come down with the diseases based on their genetics. Everyone agrees that at least five percent of all Parkinson's cases are inherited. We've identified the gene that causes an inherited form of Alzheimer's. The diseases tend to run in families. The scientists who subscribe to the genetic argument believe it's only a matter of time before we identify all the genetic markers for the diseases. Once we do, they feel, we're well on our way to curing them. It's impossible to argue with their logic.

A second group of scientists believe just as firmly that some pathogen (a microscopic critter of some sort— virus, bacteria, fungus, etc.) causes these diseases. We just haven't identified it yet. These guys have history on their side. Over and over again we've discovered that a pathogen caused a mysterious ailment rather than bad air, spirits, curses, demons, allergies or the patient's imagination. All the experts believed ulcers were caused by food, smoking or stress. Then Barry J. Marshall and J. Robin Warren discovered that the *Helicobacter pylori* bacteria did cause some ulcers; in fact, many of them. They won the 2005 Nobel Prize in Physiology or Medicine for this discovery. The idea of a virus causing cancer was considered outlandish until we discovered that viruses do cause nearly all cases of some kinds of cancer. The pathogen camp includes some very smart guys. Gregory Cochran believed that most of the chronic diseases of aging are actually caused by some unknown pathogen. Paul Ewald advanced the idea with books embraced by an army of followers who are both vocal and persuasive.

But a pathogen causing Alzheimer's or ALS? That still seems outlandish. When invaded, our bodies react by manufacturing white blood cells, developing a fever, or creating histamines (which make our noses run, for example). People with ALS or Alzheimer's don't have these symptoms. Their bodies don't try to defend themselves the way we expect them to, so we don't think a bacteria or virus is to blame. We haven't identified any living pathogen that causes the symptoms. The diseases don't seem to be "catching." We have many good reasons to assume that no bacteria, fungus or virus causes these diseases.

Before we break out the pitchforks and torches, it only seems fair to point out that our bodies ignore nearly all of the 2,000 varieties of bacteria that live within us all the time. They don't induce a fever or make our noses run. If our bodies don't manufacture antibodies for them and they don't give us a fever, science doesn't have a good way to prove we're infected. Some parasites manage to invade us and elude detection. It doesn't seem farfetched that a bacteria could take up residence within us and remain invisible to our defenses even while it pumps out toxins.

One might argue against the pathogen idea by pointing to the long latency period for lytico-bodig. What sort of pathogen hides for decades before attacking? Well, the chicken pox virus, for one. After the child recovers, the surviving virus can hide within efferent nerve cells (one-way nerves leading to muscles or glands), lurking for five decades or more until they work their way down the nerves to the skin and cause the disease "shingles." A child exposed to the virus on the skin of a shingles patient (like their grandfather) does not come down with shingles. She comes down with chicken pox. Some viruses obviously do hide and wait for decades.

Another argument in favor of a pathogen is that one disease seems to have caused a major outbreak of Parkinson's early in the Twentieth Century. About the time of a great flu epidemic, people caught a disease called *encephalitis lethargica*. They had all the symptoms of a disease caused by a virus: fever, sore throat, headache, lethargy. When that phase of the disease passed, they were left with was the symptoms of severe Parkinson's Disease, and they remained in that state for decades, cut off from the world by their condition. Oliver Sachs treated them with L-dopa and they recovered, at least for a while. He described the events in his first book, *Awakenings.*

The third group of scientists believe (just as fervently) that specific toxins cause these diseases. Their first argument is that we absolutely know that some toxins can cause the symptoms of these diseases. Some man-made toxins cause Parkinson's every single time. BMAA can cause the symptoms of ALS and was found in the brains of people who died from Alzheimer's. Rotenone can cause Parkinson's symptoms. The fungicide maneb seems to enhance the dangerous effects of paraquat or rotenone. This argument is just as compelling as the others: we know a toxin is sometimes involved, perhaps a toxin is always involved.

The toxin group has one more argument: the chemical processes involved with these diseases look an awful lot like a human body's reaction to poisons. Things like excitotoxins, genotoxins, superoxide, and inflammation keep coming up in the study of neurological diseases. They also keep coming up when you talk about toxins.

I suspect that each of these groups is partly correct. Toxins are probably involved, at least some of the time. These toxins may come from a pathogen, whether living within the patient or living in his water or food supply. Or, the toxin could be some sort of pesticide or herbicide. There is probably a genetic component to these diseases. And maybe not just the genetics you got from your parents. Peter Spencer's recent work indicates that some of the toxins the Chamorros were exposed to can affect a person's DNA and RNA. That is, a toxin can distort a nerve cell's genetic parts, even though the cell doesn't divide.

Some toxins are more potent when combined; they might also be more potent when combined with a deficiency in the defense mechanism. Genetics, pathogens, toxins, and weakened defense could all be contributing factors.

How to End a Book Like This

I wanted to end this project by saying, "... and so, to cure this disease, this is what you do ..." But I can't quite do that yet. What I'm left with is a bunch of questions. If you're just beginning your own study, I hope I've speeded up your introduction. Maybe my meanderings will spark a connection in your brain. Few of us become Isaac Newton. I'd be happy to become the apple that fell from Isaac Newton's tree, the one that started him thinking about gravity in the first place. Maybe you'll become the Newton of our time.

We started this little story with cyanobacteria, hoping we were going somewhere. Chasing clues that seemed to lead nowhere, we wandered past bats, pigs, ferns, and catfish. We noticed sick people of many kinds, quirky scientists and odd discoveries, spices and vitamins. We experienced flashes of brand-new insight that turned out to be a thousand years old. When we opened that last door and discovered cyanobacteria behind it, I knew, instinctively, that this particular journey was over for me. Learning that cyanobacteria can manufacture thiaminase felt like returning to our starting point, a natural place to conclude. I've enjoyed this project and enjoyed learning the things I did. The journey isn't over but it's time for me to step out of the maze.

A friend asked me what my "theory" was about the causes of these diseases. I told her I don't have a theory, I'm just gathering information. OK then, she insisted, what's your opinion? Good question.

It seems like these diseases strike when a person meets a perfect storm of several factors. Maybe a genetic predisposition is one factor, for example. Maybe a buildup of excitotoxins, like aspartame is a factor. The factors that most interest me are these four:

1. A toxin, which could be a herbicide, insecticide, fungicide or natural toxin like BMAA. It could be an industrial chemical. It might be a defective protein (like tau) that can reproduce itself. On the other hand, some creature (like cyanobacteria) might produce it. Maybe it's more dangerous when inhaled or injected.

2. Another toxin that acts as a catalyst, because we know that some toxins are deadlier in combinations. The fungicide maneb is an example of such a co-factor.

3. Reduced defenses (like low GST), because that would explain why some people get sick and others don't;

4. A deficiency of one or more critical minerals and/or vitamins. Iodine and thiamine would be the first suspects I'd call down to the station for further questioning. A major deficiency in either one can cause a nerve disease; maybe a minor deficiency can act as a trigger.

If a combination of those (or other) factors hit a person at once, it's easy to imagine them getting sick.

Four Impressions

I come away from this project with four overall impressions:

1. Scientists have become so focused on specialties they risk missing clues from other fields. They may have wide interests and curiosity, but one can only absorb so much information. As they research some particular enzyme and the intricate processes that create it, learning volumes of information, they simply can't also keep track of every other human enzyme, let alone what's going on in the universe of pig intestines and bracken rhizomes. What is known about fungi and bacteria is a lot less than what is unknown. If you love bacteria and a million varieties haven't even been named yet, you probably don't have the time (or desire) to study the electrical characteristics of clay mud.

One solution is to collect experts from different fields into cooperative ventures. As James Metcalf, of the Institute for Ethnomedicine says, "I am a cyano person who works with an ecologist and an ethnobotanist. We anchor a consortium of around 20 universities with specializations in Chemistry, Medicine, Neurology, Biochemistry, Pharmacology, Toxicology, etc. We also believe that bringing different disciplines together to understand diseases like ALS will give the best chance of finding cures."

Another solution is to encourage some generalists, including amateur scientists. I have renewed admiration for the offbeat, nonstandard scientists and thinkers of history — the guys like Fleming who use bacterial colonies to make little paintings, or Tesla who tried to think like the lightning thinks, or the guy who captured the pretty vapor rising from his kelp ashes, or the lady who noticed that eating different kinds of mud sometimes cured people, or the guy who drained the blood out of dogs to see what might cure them, or Lynn Margulis who promoted the idea of "endosymbiosis" which people ridiculed for decades until they decided she was correct.

It's just fine to make wild, weird leaps of the imagination. We should encourage and honor that kind of thinking because, every now and then, a bizarre idea proves useful. Sometimes, it's hard to identify the Steve Jobs of the world when they're surrounded by a mob of ordinary guys in jeans and black turtlenecks. On the other hand, we should be careful not to confuse "this is a cool idea" with "this is true."

2. The Internet preserves old information, yet old "knowledge" is becoming less familiar, not more. Everyone seems to have forgotten about "miracle drugs" and snake oil cures like iodine and lithia water. What if the hoopla wasn't merely placebo optimism? Even fascinating new studies about turmeric, iodine, hydrogen sulfide, stem cells and thiamine risk drowning in the ocean of information, much of which is simply false.

3. The interconnection of different life forms stunned me with its pervasiveness and complexity, once I focused on it. Plants don't thrive without a fungal network around their roots. Bacteria create vitamins and digest food for animals. Other bacteria create toxins that protect animals (and perhaps plants as well). Good bacteria fight with bad bacteria. Fungi fight with bacteria or join forces with algae to form lichens. Some bacteria fix nitrogen for plants. Important little organelles within animal and plant cells probably began as bacteria that invaded and stayed forever. We may never understand life as observed in a test tube, because life exists within a writhing swarm of other life. Without the natural chaos and confusion, one cell in a test tube remains forever out of context. And finally:

4. Wouldn't it be fun to be a young scientist right now, with a whole career ahead of you?

Loose Ends

As a practical matter, avoiding toxins seems like a good first step. I won't be swimming in pond scum in the near future. I won't be fishing in it. I won't be eating dairy products from my local farmer if he lets his cattle drink green water, whether he calls it organic or not. I'm going to be more careful about how many diet soft drinks I consume, because most of them contain the excitotoxin "aspartame." I'll be more careful about MSG in my diet. I'll avoid herbicides, fungicides, and insecticides as much as I can. I'll wash my tomatoes, honest I will.

I remain very curious about the role of iodine, turmeric, thiamine, lithia water, zinc, magnesium, Vitamin C, Vitamin E, and hydrogen sulfide. None of these are patentable, so studying them lacks a profit incentive. Early studies (from the 1930s and 40s) indicated that ALS results improved when drugs were coupled with large doses of Vitamins A and D.

I'd still like to know why rats and bats respond to some poisons differently than monkeys or humans.

Eating clay or charcoal to eliminate toxins seems a less goofy idea than it did when I first heard about it. I'm not ready to eat mud, but geophagy no longer seems like evidence of a mental disorder. Similarly, I have an odd new respect for people who love burned toast. Maybe it acted like charcoal to cleanse grandpa of toxins. Does the idea of studying burned toast sound a little weird to you? It does to me too, but maybe better than eating mud.

Several fortunes will be made by developing tests for toxins. Right now, we can't test for many of them on a large scale.

The balance of good bacteria and bad bacteria within our intestines seems complex, important, and not well understood. Some people suggest that conditions as different as autism and obesity are controlled by the balance of bacteria in our internal army. That microscopic jungle depends on us for food as well as for micronutrients of all kinds. I wonder how our diets affect their metabolism? Surely bacteria have their own nutritional requirements; it would probably be smart to feed our little buddies well and starve our little enemies.

Epilogue: Hope

You might read all this and feel a touch of despair that it's all too mysterious, too confusing; that we're probably decades away from curing these diseases.

I come away with the opposite feeling. I've been trying to make sense of this in an old fashioned, nineteenth century kind of way, just by reading. I've been flying over it all (probably in a colorful and highly decorated hot air balloon contraption) trying to understand the forest thousands of feet below as I glimpse treetops through the cloud cover. But, rest assured, an army of well trained scientists are on the ground, ankle deep in the reality of the diseases and conducting experiments on a microscopic level. Every day, they discover new proteins and chemicals and conduct exhaustive tests. They argue with each other, they hold conferences, they repeat experiments and write detailed reports using words I've never heard and can't pronounce. One single discovery could catapult our understanding ahead in a blazing leap. It could answer all the questions; it could happen any day.

I've only talked about the tip of the iceberg. I haven't mentioned stem cell research, for example, or interesting chemical discoveries, mostly because those involve harder science and foreign languages.

Stem cell research sounds very promising. Stem cells can be directed to grow into cells of just about any variety. If we could replace damaged nerve cells with brand-new ones, we might reduce these diseases to footnotes in future history books. Researchers can now create stem cells from a patient's own skin cells, eliminating most ethical and religious questions. They've sped up the process dramatically. This isn't science fiction; it's already widely used in bone marrow transplants.

Lee Martin, a pathologist at Johns Hopkins University School of Medicine has transplanted stem cells into mouse brains and delayed the onset of ALS. That sentence alone should provide hope.

Back in the olden days of 2004, China was already experimenting with implanting fetal stem cells into the brains of patients with ALS and showing nearly immediate improvement. I don't know how far their research has progressed, but the seven or eight years since that report is the equivalent of a hundred years of nineteenth century science. Because the U.S. government put strong restrictions on stem cell research in 2001, I'd have to learn French and Chinese to be able to read most of the studies since then. Now that U.S. scientists can investigate stem cells more freely, we can expect an avalanche of new information to be published in English very soon.

In 2014, Su-Chung Zhang, a neuroscientist at the Waisman Center at the University of Wisconsin and his colleague Hong Chen discovered a fascinating clue to the mechanism of ALS. Three specific proteins are critical to the operation of neurofilaments, structures which carry vital chemicals within nerves. According to Zhang, one faulty gene can cause misregulation of these neurofilaments, which causes tangles. Tangled neurofilaments don't function well, if at all. Zhang and Chen also noted that similar tangles occur in patients with Alzheimer's and Parkison's, not just ALS. They don't know what causes the gene to become defective, because they're studying the mechanism rather than the cause. Interestingly, when they "edited" the gene responsible for the tangles, the cells began to function normally again. If this work continues to be as promising as it seems right now, it may not matter what causes these diseases— we might be able to repair the damage, even if a dozen different factors led to it in the first place. It would be like cooling a cup of hot water: it doesn't matter if you heated it in the microwave, or on your stove, or over a campfire. Throwing an ice cube into it will cool it down.

In May, 2014 Science News reported on studies by a team lead by Academy Research Fellow Jaan-Olle Andressoo at the Institute of Biotechnology in University of Helsinki, Finland. The article said: "MicroRNA-206 was in 2012 shown to down-regulate Brain Derived Neurotrophic Factor (BDNF) levels. When in the animal model of Alzheimer's disease the activity of microRNA-206 was blocked, BDNF levels rose and Alzheimer's disease features were alleviated."

Several simple nutritional ideas seem promising and you might want to investigate them further. The nerve cells of Alzheimer's patients don't seem receptive to glucose. Some people think this is critical. They point out that those brain cells remain receptive to other kinds of fuel, like ketones. By consuming more coconut oil, some people claim to have improved the brain function. My sister reports that a good friend of hers had near-miraculous results by feeding her mother, an Alzheimer's patient, a spoonful of coconut oil every day for three weeks. The patient spoke for the first time in two years and sat up by herself. Stories like that are intriguing, but not scientific. Could be complete nonsense.

The role of a specific protein called "tau" has always been interesting to people studying Alzheimer's, but the details have eluded us. In early 2012 scientists reported new insights into this. They now believe that abnormal tau can spread from cell to cell within the brain almost like an infection. Discovering this could lead to dramatic and sudden new methods for reversing the disease.

In 2014, scientists at Harvard including, Amy Wagers, published two studies which indicate that the blood of young mice can reverse many of the symptoms of aging in old mice, including memory loss. Interestingly, in a reverse experiment, the blood of old mice tended to age young mice. They believe they've isolated the protein responsible for this and will do further research. A group from Stanford and the University of California at San Francisco, including Tony Wyss-Coray, published similar findings at the same time, although the teams weren't working together. The Stanford group intends to start human trials on Alzheimer's patients nearly immediately.

In 2014, scientists at the University of Utah (led by Denise Dearing and Kevin Kohl) discovered that pack rats can eat toxic creosote plants and juniper (which they love) only if their intestines contain specific bacteria. If you kill those bacteria the rats can't deal with the toxins and begin to waste away, despite continuing to eat the plants. Implanting specific bacteria into their intestines "cures" them. Think about that. At least in pack rats, specific bacterial colonies in the gut provide some sort of defense against certain poisons. If a toxin causes a nerve disease in humans, and one of the body's defenses against that toxin is a bacteria, what happens when the bacterial population becomes unbalanced? I wonder if future doctors will treat diseases by adjusting the mix of bacteria in our intestines?

Each of the following phrases (which you can google) represents new science that would be worthy of its own book: angiogenin; gelsolin; "broken protein recycling process" melittin; and the role of uric acid.

Excellent scientists are racing against each other, on many different paths, to solve the puzzle. The winners stand to make a lot of money so their scientific curiosity is amplified by a financial incentive. I would not bet against them.

Another reason to be hopeful is this book. Not the book itself (I don't have *that* big an ego) but that the fact that even a guy like me, starting from near zero knowledge, could assemble all this information in a few months using only my suburban library, a simple computer, and an Internet connection. A very few years ago that would not have been possible. I've been able to download and read books from a hundred years ago as well as studies published yesterday. On a much grander scale, scientists and doctors all around the world now routinely use computers and the Internet to make connections, collect data, and compare notes. They have access to much more information than you or I do. We've seen how the acceleration of technology has transformed every area of our lives. Within ten years we moved from a world without i-pods to a world where children carry cell phones more powerful than any computer that existed when they were born. That same transformation is also gaining speed in the life sciences.

New discoveries made tomorrow will circle the globe by day after tomorrow. Any one of them might change your life, or mine, or Suzanne's.

About the Author

Kenn Amdahl is best known for writing "funny books on dull subjects." His nonfiction books include *There Are No Electrons: Electronics for Earthlings; Joy Writing: Discover and Develop Your Creative Voice; The Wordguise Alembic Volume One; Algebra Unplugged* (with Jim Loats, Ph. D.); and *Calculus for Cats* (with Jim Loats, Ph. D.). His novels include *The Land of Debris and the Home of Alfredo*; and *Jumper and the Bones*. His books have sold hundreds of thousands of copies and are credited with inspiring all the books for Dummies and Idiots.

In 1990 he formed Clearwater Publishing Company to publish his own books and became a pioneer in the micro publishing industry. He still publishes his own books.

He's been active in the Colorado book community ever since he began thinking of himself as a writer and has served on the board of directors of over a dozen writing and publishing associations. His poetry has been published in literary journals in nearly every state; his songs have been played on over a hundred radio stations around the world.

He lives with his wife of over 40 years in Broomfield, Colorado. They have three grown sons.

Resources

Here are some websites that may be useful for learning more about these diseases. The websites here and in the "reference" page and acknowledgement page look long because, in the electronic version of this book, they are live links directly to each site.

We have no relationship with any of them and no direct knowledge about their reputability:

ALS Association: http://www.alsa.org/

ALS Foundation for Life: http://www.alsfoundation.org/

The Alzheimer's Association: http://www.alz.org/

American Parkinson's Disease Association: http://www.apdaparkinson.org

Parkinson's Foundation: http://www.parkinson.org/

Multiple Sclerosis Association of America: http://www.msaa.com/

National Multiple Sclerosis Society: http://www.nationalmssociety.org

Oregon Health Sciences University: http://www.ohsu.edu/

Linus Pauling Institute at Oregon State University: http://lpi.oregonstate.edu/

The Institute for EthnoMedicine: http://208.106.229.203/about/default.asp

Weil Cornell Medical College: http://www.med.cornell.edu

Ikaria Pharmaceuticals: http://www.ikaria.com

Preventative Medicine Group: http://www.prevmedgroup.com

GreenMed Info.com: http://www.greenmedinfo.com

References

In the text of this book, I've mentioned several books I read in their entirety. Some especially interesting ones include:

At Home, by Bill Bryson

http://www.amazon.com/At-Home-Short-History-Private/
dp/0767919394/ref=sr_1_1?s=books&ie=UTF8&qid=1334783360&
sr=1-1

Cycad Island, by Oliver Sachs

http://www.amazon.com/The-Island-Colorblind-Cycad-Signed/dp/
B001GTZD2Y/ref=sr_1_4?s=books&ie=UTF8&qid=1334783261&sr=1-4

The Biology of Bats, by Gerhard Neuweiler

http://www.amazon.com/Biology-Bats-Gerhard-Neuweiler/
dp/0195099508/ref=sr_1_1?s=books&ie=UTF8&qid=1334783093&
sr=1-1

For Love of Insects, by Thomas Eisner

http://www.amazon.com/For-Love-Insects-Thomas-Eisner/
dp/0674018273/ref=sr_1_1?s=books&ie=UTF8&qid=1334783390&
sr=1-1

Growing Gourmet and Medicinal Mushrooms, by Paul Stamets

http://www.amazon.com/Growing-Gourmet-Medicinal-Mush-
rooms-Stamets/dp/1580081754/ref=sr_1_1?s=books&ie=UTF8&qid=1334
783419&sr=1-1

Growing Mushrooms the Easy Way, by Rush Wayne

http://www.amazon.com/Growing-Mushrooms-Easy-Way-Cultiva-
tion/dp/B0041OWTG4/ref=sr_1_1?s=books&ie=UTF8&qid=1334783518
&sr=1-1

The Iodine Trail, by John Stanbury

http://www.amazon.com/The-Iodine-Trail-Deficiency-Prevention/
dp/0195698770/ref=sr_1_3?s=books&ie=UTF8&qid=1334783484&
sr=1-3

Island Bats, by Theodore H. Fleming, Paul A. Racey

http://www.amazon.com/Island-Bats-Evolution-Ecology-Conservation/dp/0226253309/ref=sr_1_1?ie=UTF8&qid=1334783059&sr=8-1

Island of the Colorblind, by Oliver Sachs

http://www.amazon.com/The-Island-Colorblind-Oliver-Sacks/dp/0375700730/ref=sr_1_1?s=books&ie=UTF8&qid=1334783195&sr=1-1

Mycelium Running, by Paul Stamets

http://www.amazon.com/Mycelium-Running-Mushrooms-Help-World/dp/1580085792/ref=sr_1_1?s=books&ie=UTF8&qid=1334783453&sr=1-1

Tomatoland, by Barry Estabrook

http://www.amazon.com/Tomatoland-Industrial-Agriculture-Destroyed-Alluring/dp/1449423450/ref=sr_1_1?s=books&ie=UTF8&qid=1334783327&sr=1-1

It seems wasteful to use paper and ink to print a huge bibliography that only a tiny fraction of readers will even peruse. Therefore, I intended to create a web page with as much of that information as is practical. It will be available at http://www.clearwaterpublishing.com/2010/revenge-of-the-pond-scum-references/ So far, that task has proved more formidable than my determination to do it. On the other hand, I kept "notes" of interesting paragraphs emails, and articles I found online by copying them and pasting them into one long text document which I later studied by using a global search for terms. That document is about 250,000 words long, or more than three times as long as this book. If you really, really want to study that, and you promise not to plagerize anyone else's words, I could email you a copy of that.

If you'd like that, or want help locating anything mentioned in this book please email the author at Wordguise@aol.com

Acknowledgments, with websites

Thanks to:

Paul Alan Cox, Ph. D. (founder and co-director of the Institute for Ethnomedicine)

> http://www.ethnomedicine.org/about/pcox.asp

Cynthia Dormer, Ph.D. (associate professor of nutrition, Metropolitan State University of Denver)

> http://www.mscd.edu/searchchannel/jsp/directoryprofile/profile.jsp?uName=cdormer

Balz Frei, Ph.D. (director of the Linus Pauling Institute at Oregon State University and expert on exitotoxins and superoxide)

http://lpi.oregonstate.edu/staff/freibio.html

Stephen Lawson, (Linus Pauling Institute Administrative Officer, biographer of Linus Pauling)

> http://lpi.oregonstate.edu/

Derrick Lonsdale, MD (thiamine expert, Cleveland Clinic, Fellow of the American College of Nutrition (FACN), Fellow of the American College for Advancement in Medicine (FACAM)

> http://www.prevmedgroup.com/lonsdale.php

James Metcalf, Ph.D (co-director of the Institute for Ethnomedicine, cyanobacteria expert)

> http://www.ethnomedicine.org/about/jmetcalf.asp

Rob Mies, Ph.D. (founder and director of the Organization for Bat Conservation at Cranbrook Institute)

> http://www.batconservation.org/drupal/director-bio

Lisa Miller, (Brookhaven Institute, photographer)

> http://infrared.nsls.bnl.gov/u10b/research/people.htm

Peter Spencer, Ph.D (director, Oregon Health Science Universities Global Health Centers, expert on BMAA)

http://www.ohsu.edu/xd/research/centers-institutes/croet/faculty/profiles.cfm?facultyID=520

Rush Wayne, Ph.D. (author and expert on fungi)

http://www.mycomasters.com/About-author.html

John Stanbury, Ph.D. (Founder of ICC-IDD, iodine expert)

http://www.iccidd.org/

http://www.theiodinetrail.com/book

Photographs on the cover of E book and/or printed book:

"microcy10" (the image behind the title and author's name) John Patchett (University of Warwick), Mark Schneegurt (Wichita State University), and Cyanosite (http://www-cyanosite.bio.purdue.edu)

"nostoc2c" (the image behind the subtitle) Roger Burks (University of California at Riverside), Mark Schneegurt (Wichita State University), and Cyanosite (http://www-cyanosite.bio.purdue.edu)

Gerry Morrell and Morrell Printing: http://morrellprinting.com

Liz Hill   http://www.lizhill.net

Karen Reddick: http://www.TheRedPenEditor.com

Gary E. Gibson, Ph.D., Weill College, Cornell University

http://vivo.med.cornell.edu/display/cwid-ggibson

Betsy Dumont, Ph. D., UMass Amherst

http://www.bio.umass.edu/biology/about/directories/faculty/elizabeth-r-dumont